TRIBUTARY VOICES

Waterscapes—History, Cultures, and Controversies

William Rowley (emeritus, University of Nevada, Reno), and
Paul Lindholdt (Eastern Washington University)
Series Editors

At the most fundamental level, water has always been essential to life. Yet it is becoming an increasingly contentious resource—from how to allocate the Colorado River in the American West to drought and coastal flooding the world over. In some places there is too little, in others too much, and future projections are uncertain at best, dire at worst. With its geopolitical implications, water is the oil of the twenty-first century. Climate change is a related challenge that affects the entire world, indiscriminately and with devastating and wide-ranging results. Pollution and water quality continue to be problems not just in the developing world but domestically, as Flint, Michigan, reminds us. Water has also played a cultural role in the human imagination throughout history. Most ancient traditions mention a great flood. And what would Thoreau be without his pond?

Waterscapes is an interdisciplinary series in which we recognize the centrality of water in the past, present, and future. We recognize water in a very tangible and practical sense, as well as a spiritual and reflective sense. We welcome approaches from the social sciences and humanities, including ecology, ecocriticism, nature writing, travel, history, political science and public policy, literary and cultural studies, and international relations. Single-site and comparative works, and works of global scale will be included, as will a variety of methodologies.

Tributary Voices: Literary and Rhetorical Explorations of the Colorado River
Paul A. Formisano

Contents

Preface

Some of us who live in arid parts of the world think about water with a reverence others might find excessive.
—Joan Didion, *The White Album*

This is a story about a river: the Colorado. Yet unlike the many source-to-sea narratives that chronicle adventurers' journeys from majestic, verdant mountainscapes through canyon country to the burning deserts of the Lower Basin and the eventual delta, the work before you is more of a journey among words and ideas. While the Colorado River and the basin it drains are of utmost importance to this study, my focus addresses how various stakeholders have imagined this river and its diverse topographies over time and how these imaginations and their respective expressions have given rise to the ways in which the river and watershed have been and are currently shaped. My goal is to examine these imaginations for what they might teach us about our relationship to the river and how we might—through the words and ideas of those addressed in the present study—enact a more sustainable relationship with the Colorado.

In many ways, however, this book is also my story of the peoples and places I have encountered over the past four and a half decades. I was raised along Colorado's Front Range with a childhood full of outdoor pursuits. There, on the rolling hills south of Denver, I marked my sense of place and belonging by noting the three sentinels of the Front Range: Pikes Peak, Mount Evans, and Longs Peak. Each mountain rising more than fourteen thousand feet above sea level, their snow-clad slopes greeted me each day on my way to school and offered numerous opportunities growing up to immerse myself in the southern Rockies' high alpine through many trips with the Boy Scouts, church groups, and friends and relatives. With family living on Colorado's Western Slope, many of my earliest memories involved road trips in the back of a station wagon, the car winding its way up Interstate 70 where it crossed the Continental Divide through the Eisenhower Tunnel. Passing Colorado's famed ski resorts, we continued westward as mountain peaks gave way to mesas and plateaus. There the Colorado River emerged from the north at the tiny town of Dotsero to run alongside the freeway through the magnificent Glenwood Canyon, where we spent many hours sitting idle on the interstate watching the canyon's incredible freeway-expansion project develop. From there, the river and freeway sliced through the sandstone cliffs of De Beque Canyon and into the Grand Valley, our final destination.

What I didn't know at the time on all of those journeys west was how much the Colorado River system would influence my life. Later as an undergraduate in Utah, I worked for a summer as a field assistant to a PhD candidate studying riparian areas along the dammed rivers and creeks of the Uintah Basin. There, I spent long days bushwhacking back and forth across a drainage to set up transects to map the plants and groundwater below the dams, with evenings spent lounging in camp, reading, spotting the occasional moose, and listening to the wind blow through the sage. As my academic interests and strengths eventually led me to the humanities where I was first exposed to environmental literature, I sought out opportunities to blend my growing knowledge of this field with my experiences on the ground.

Shortly after graduation, and prior to enrolling in the Literature and Environment graduate program at the University of Nevada, Reno, I applied to participate in International Project WET's "Discover a Watershed: The Colorado" project to promote water literacy by developing K–12 curriculum alongside educators from throughout the watershed. With the great fortune of being selected to participate, I joined these teachers and a small group of college students from other western and Mexican universities in workshops about the river. Months later, we students gathered with our project leads for a six-week journey through the watershed to meet with numerous river stakeholders and learn about their approaches to living with the Colorado. Starting in Grand Lake, Colorado, we traveled to the headwaters of the Green River in Wyoming, then south along the Green into Utah, through the Grand Canyon by raft, and on to Mexico toward the river's terminus. Never having seen the river below the Utah-Arizona border, I was struck by the changes the river had undergone in meeting the demands of Southern California's and Arizona's productive agricultural valleys. Yet Mexico was most eye-opening, as the river stopped flowing in its traditional channel below Morelos Dam. While the rapids of the Grand Canyon and the high country of Colorado and Wyoming were perhaps the most entertaining and aesthetically impressive parts of the tour, the Ciénega de Santa Clara, an extensive wetland in northern Sonora created by Arizona's agricultural wastewater, was the most influential on my thinking. Here I pieced together the entire river and its larger watershed to see the material impacts of nearly forty million people who depend on the river. This was an education unlike any other.

A subsequent internship with the Colorado Foundation for Water Education (now Water Education Colorado), graduate work in Nevada and New Mexico, and a National Endowment for the Humanities Summer Institute on Hoover Dam reinforced my fascination for and commitment to the river, its watershed, and diverse cultures. Looking back on my lifelong connection to the watershed, I realize now how, in many ways, the genesis of *Tributary Voices* mirrors my own awakening to the watershed's many concerns. Whereas recreational pursuits defined much of my earliest interests in the region, years of traveling throughout the watershed, meeting with many different stakeholders, and reading, writing, and teaching about the Colorado have greatly enriched my understanding of the river's complexity and alerted me to the great disparities and environmental justice issues that exist throughout the basin. They have reminded me of my privileged upbringing and instilled in me a desire to work toward a more equitable and sustainable Colorado for my children and for the generations to come who will look to this remarkable watercourse for survival. Thus, *Tributary Voices* is in so many ways a confluence of the various streams of my own life, a humble attempt to consider alternative solutions and perspectives to those that have long framed discussions about water conservation.

This journey includes a rich cast of characters who have influenced this project and supported it from the beginning. To Bruce Glisson I'm grateful for trusting a nonscientist to assist in his fieldwork. I have fond memories of walking for hours along and through the Uintah Basin's rivers and streams. To all those involved with the 2003 Discover a Watershed: The Colorado project—Justin and Lissa Howe, Valerie Gates, Linda Hveem, and Dennis Nelson as the project's leaders, organizers, and inspiration; Mathieu Brown, Martha Evers, Jacob Fillion, Andre Potochnik, and R. J. Johnson for guiding us safely through the Grand Canyon and revealing the endless wonders of that remarkable place; and to the rest of the crew, Larissa Conte, Maria Goodin, Karen Hyun, Craig Maier, Chris Newell, Barbara Peralta, Carlota Monroy, Mauricio Torres, and Paul Viosca—you gave me one of the greatest summers of my life and provided the initial impressions that would eventually coalesce into this book. From freezing in the Wind Rivers to boiling in the delta, I couldn't have been with a more fun, engaged, and inspiring group of people. I thank Karla Brown for giving me an opportunity to rub shoulders with Colorado's water community, slog through muddy spring pastures along the Little Snake, and paddle the Yampa. Thanks to Justice Greg Hobbs for sharing with me his love of poetry and green chile as well as introducing me to C. F. Blanchard's writings on reclamation.

This project would not exist without superb academic mentors over many years who patiently entertained my ideas, astutely critiqued and improved my writing and thinking, and have generously given of their time to support my professional development. George Handley first introduced me to the field of environmental literature and has continued, in subsequent years, to influence my thinking about humanity's relationship with the earth. Brian Cannon and Tom Alexander helped me better understand the uniqueness of the West's history and environment. Ann Ronald endured my attempts to write creatively about the river and gave much-appreciated advice during my master's program.

Michelle Kells's influence is felt throughout the book through her commitment to environmental justice and for helping me articulate a discursive ecology of water literacy. To the participants of the National Consortium of Environmental Rhetoric and Writing's inaugural workshop in Taos, thank you for your help in reshaping the book's introduction. This book would not exist without the guidance and encouragement over many years from Gary Harrison. He reminded me of the value of this work at a time when I doubted its relevance. May you enjoy many hours on the trail and laughter with your family.

My colleagues at the Western Literature Association have provided valuable insights over the years, enriching my love of the North American West and creating wonderful conferences to look forward to each fall. Anthony Arrigo, Michael Green, and the participants of the 2018 National Endowment for the Humanities Summer Institute on Hoover Dam provided great conversation and insights about western water issues during our two-week sojourn in Boulder City, Nevada. We were quite a formidable presence at the local trivia night.

For the many pleasant hours spent in the archives, I wish to thank the library staff at Brigham Young University; the Church of Jesus Christ of Latter-day Saints' Church History Library; the University of Nevada, Las Vegas; the University of New Mexico; the University of Utah; and the Utah State Archives. Tom Arviso, Duane Beyal, Jeanette Chapman, Ashley Davidson, Danielle Grabish, Tomas Jaehn, Kristal Kraft, Jen Kirk, Joaquín Murrieta-Saldivar, Deidre Peaches, Kristine Priddy, and Liz Rogers were particularly helpful in helping locate images and granting permissions.

The University of Denver's Sturm College of Law's 2018 Symposium and Colorado Mesa University's Hutchins Water Center's 2020 Water Course have reiterated the importance of sound science, just laws and policy, and the insight the environmental humanities bring to our water challenges.

My fellowship with the Charles Redd Center for Western Studies at Brigham Young University was indispensable to this book's completion. Notwithstanding a global pandemic, Jay Buckley, Brenden Rensink, and Amy Carlin ensured a highly productive and refreshing semester. The breathtaking views of the Wasatch from my office window inspired the writing and aided my thinking about my own connections to place and life in the arid West. An earlier research grant from the Redd Center enabled archival research on Navajo Dam. The center is a great patron of western American scholarship, and I will be forever grateful for my time there.

To Darlene Farabee and the rest of my colleagues in the University of South Dakota English Department, I owe you my most sincere thanks for making my time there so pleasant. Such sentiments aren't always felt within the halls of academia, and I've greatly valued the ongoing support and camaraderie over the past (almost!) decade. My affiliation with USD's Sustainability Studies Program and opportunities to present to students in those courses have reflected the type of interdisciplinary work that needs to occur more often across college campuses. The College of Arts and Sciences has been very generous in supporting travel and research over many years.

I am indebted to Margaret Dalrymple, JoAnne Banducci, and Caddie Dufurrena at the University of Nevada Press and the manuscript's reviewers for all their work in guiding me through the book's production stage and marketing.

Finally, my family deserves my most heartfelt gratitude. My in-laws, Larry and LaDawn Porter, hosted us during my sabbatical and made sure no one went hungry. To the rest of the Porter clan, you are fantastic. Lynn and Deb, your love of the West and knowledge of water helped me get my feet wet. My parents, Howard and Diane, first introduced me to the river, and with them, my siblings, and their families, I have spent many wonderful hours immersed in nature's beauty. Shanna, Olivia, Maren, James, and Anne—you are my bedrock. Thank you for holding on during this wild ride and for always being ready to adventure near and far. It is to you that this book is dedicated.

TRIBUTARY VOICES

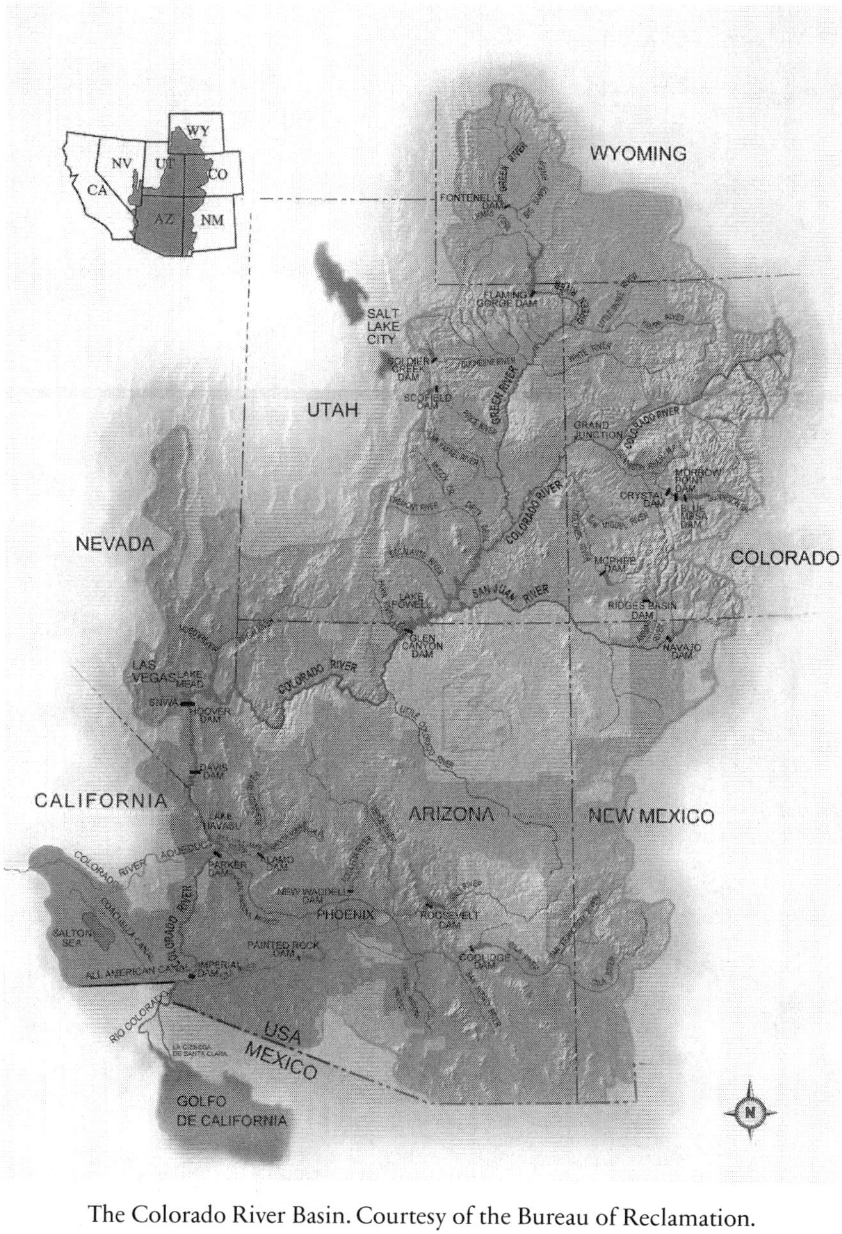

The Colorado River Basin. Courtesy of the Bureau of Reclamation.

Introduction

A River of Words

There is a unity about the Colorado River country that is established by the River itself—always the River.
—Francis P. Farquhar, *The Books of the Colorado River and Grand Canyon*

The Colorado River is the lifeline for the nearly forty million people who reside in seven US states in the American West and two in northern Mexico.[1] Its headwaters begin in Wyoming's Wind River Mountains and Colorado's Rocky Mountain National Park, the water trickling from snowmelt to create small streams and creeks that eventually form the Green and Colorado Rivers, which merge in the canyon country of southeastern Utah. The unified Colorado then tumbles onward through magnificent canyons including Glen, Marble, and the Grand to the gentler but more arid topography of the Lower Basin. There the river flows south, forming the boundary between California and Arizona toward its historic exit in the Gulf of California. Covering roughly 242,000 acres in the United States and 2,000 more in Mexico, the Colorado River Basin is a land of extreme diversity and beauty. For thousands of years it has inspired fear, reverence, awe, and desire in those who have experienced this region. Such emotions continue to shape perceptions about the basin and river, particularly as the latter faces unsustainable demands of its limited resources.

Today, the Colorado is a far cry from the historically unpredictable, silt-laden river that the region's original inhabitants and subsequent explorers would have known. Although the river maintains some of its former luster, it has been described as "a river no more," as thousands of ditches, dams, and headgates divert much of the Colorado's flow toward exploding urban centers and expansive agricultural areas (Fradkin). Throughout the Colorado Basin, this infrastructure impedes countless creeks, streams, and rivers, with the largest projects like Blue Mesa, Flaming Gorge, Glen Canyon, Hoover, and Navajo Dams pooling the waters behind their concrete facades for hundreds of miles. Depending on one's point of view, these dams have brought countless economic benefits or environmental disasters to the arid West. They have controlled massive floods that once ravaged the Lower Basin and brought water to places like California's Imperial Valley, one of the most fertile regions on earth. The dams' reservoirs also constitute some of the West's most coveted recreation sites and have allowed cities like Los Angeles, Las Vegas, Phoenix, Salt Lake City, Denver, and Albuquerque to flourish by generating hydroelectric power and providing much-needed water to slake their growing thirst.

Yet the reclamation projects' impacts on the environment have been equally dramatic. They have flooded places like Glen Canyon, noted by John Wesley Powell and other explorer-adventurers as one of the most remarkable places on earth, or, in Edward Abbey's words, "a portion of earth's original paradise" (*Desert Solitaire* 189). They have caused irreparable loss of native fishes and wildlife habitat; increased harmful levels of bacteria, chemicals, salt, and hard metals in many of the reservoirs and surrounding agricultural lands; and led to the widespread damage to the delta region and its human and nonhuman communities alike—a place that one writer has argued, "Only the Nile and the Indus Rivers were comparable for spectacle and scale" (Bergman 57). Once a region of nearly two million acres—the size of Rhode Island—that supported millions of birds like the Yuma clapper rail and other species like jaguars, the *vaquita*, and the clam *Mulinia coloradoensis*, which have either disappeared from the region or are listed as endangered, the delta is only a shadow of its former splendor (Luecke et al. 2). The countless demands on the river by agricultural, industrial, municipal, and recreational sectors have stretched the river to the limit. In fact, such increasing demands and persistent drought throughout the region have prevented the river from reaching the Gulf. More often than not, the river all but dries up at the Mexican border as farms divert the remaining water into the green fields of the Imperial and Mexicali Valleys. Only in extremely wet years when reservoirs throughout the Upper and Lower Basins release surplus water or when a binational restoration experiment occurs like that in 2014 does the river reach its historic outlet in the Gulf of California, 120 miles south of the US-Mexico border.[2]

For most of the twenty-first century, the Colorado River Basin has been mired in "a historic, extended drought," leaving reservoir levels far below their full capacity ("Drought"). In fact, during the period between 2000 and 2004, the Colorado experienced its lowest five-year average for flows, while Lake Mead and Lake Powell lost 50 percent of their overall capacity (*Draft Report* 2). By 2009, researchers had concluded that a 20 percent reduction in available river water caused by climate change coupled with increased demands and current management strategies could reduce available storage by 51 percent by 2057 (Rajagopalan et al. 4). Over the past decade, discussion in the water community has increasingly become preoccupied with this ever-dwindling storage at Lake Mead and the increasing possibility that the federal government will issue a shortage declaration for the river, a move that would have immediate impacts on Arizona's and Nevada's Colorado River allocation.[3] This looming reality, along with the ongoing drought and increased demands, has earned the Colorado and shorter stretches of the river the label of "most endangered" river by the nonprofit American Rivers in 2013, 2015, and 2017 (Owen 57–58; Kober). In 2018 scientists concluded that the US West has experienced a 21 percent reduction in snowpack since 1915 and that warming trends will continue to reduce the necessary runoff that keeps the arid West alive (Mote et al. 2). This is despite such periodic spikes in the river's hydrograph evident in above-average snowfall such as those during the 2016–17 and 2018–19 winters that have postponed what seems like the inevitable given current trends (Brean). Today, scientists use the terms "megadrought" and "hot drought" to describe the basin's climatic reality—terms whose implications should cause significant pause for those dependent on the Colorado (Williams et. al; Udall and Overpeck).[4] These realities have caused the Colorado Basin's major reservoirs to continue their decline through the 2021 summer, leading the federal government to issue its first-ever shortage declaration for the Colorado on August 16. As of the first week of October 2021, the beginning of the basin's water year, Lake Powell was only at 30 percent capacity, while Lake Mead stood at 35 percent, the latter dropping to its lowest level ever recorded at 1,067.35 feet (Lower Colorado; "Accumulations"). Indeed, we are entering uncharted waters when it comes to Colorado River management.

Such dire conditions have encouraged numerous agencies throughout the watershed to work together to better grasp the implications of widespread drought, which suggests a new normal for the river. In 2007, the secretary of the interior released its *Record of Decision*, also known as the Interim Guidelines, which outline the seven basin states' plan to promote conservation throughout the region and establish the joint operation of Lake Mead and Lake Powell to better respond to future reductions in river flow.[5] To plan for such potential shortages, the Bureau of Reclamation released its comprehensive Colorado River Basin Water Supply and Demand Study in 2012. The region's tribal nations followed suit, partnering with the bureau to publish the 2018 Colorado River Tribal Water Study, which will be addressed in greater detail in chapter 5. In 2019, the basin states, along with some of the major agricultural and municipal water users in the Lower Basin, presented the Colorado River Drought Contingency Plan Authorization Act to Congress to respond to the growing uncertainty caused by drought and climate change since the signing of the 2007 Interim Guidelines. This act outlines the participants' plans to manage Lake Powell and Lake Mead to spread the burden of drought more evenly throughout the basin and "to avoid circumstances which could otherwise form the basis for claims or controversies over interpretation or implementation of the Colorado River Compact and other applicable provisions of the Law of the River"—the extensive body of laws that govern river management between states, tribes, and nations (*Drought Contingency* 3). Most recently, the Western Water Assessment compiled its "Colorado River Basin Climate and Hydrology: State of the Science" report in April 2020 to synthesize the various tools and data sets used to generate the region's near- and long-term forecasts. Such commendable efforts across so many different organizations speak to a growing acceptance that "patterns of water use that became established (even entrenched) during the climate of the past cannot be changed without intense political effort owing to large cultural, economic, and infrastructure investments in the status quo ante" (Mote et al. 4).

While these various reports largely address issues at the basin level, other initiatives over the past two decades targeted more local efforts. In response to findings that water crises were "highly likely" along Colorado's Front Range and Utah's Wasatch Front as well as in southern Nevada and the Santa Fe–Albuquerque region, the US Department of the Interior and the Bureau of Reclamation unveiled their Water 2025 program to encourage greater conservation along these urban corridors ("Water 2025").[6] Other programs such as Las Vegas's "cash for grass" initiative paid home owners to replace water-loving turf grass with xeriscaping; other cities throughout the basin have adopted similar plans (Sweet). Installation of low-flow toilets and showers, tiered water metering for watering lawns, and other initiatives have cut back on residential water use. As agriculture constitutes nearly 80 percent of all Colorado River water use, changes to irrigation practices have helped conserve water in the hottest reaches of the basin, and numerous cities and farming communities have collaborated—albeit through significant economic incentives—to fallow fields and divert water to growing metropolises (Babbitt). All of these efforts have played an important role in reducing water use and signaling that change is possible.

Science, Language, and a Crisis of Representation

Such challenges require innovation and adaptation to changing material conditions. With green lawns still draping so many western communities that rely upon the Colorado River and agricultural practices that seem far better suited for more humid climes, the dire projections embodied in the countless studies and articles may seem a far-off dystopian fiction for most living in and around the Colorado Basin—one akin to what Paolo Bacigalupi portrays in his thriller *The Water Knife* (2015). However, circumstances such as those during California's drought from 2011 to 2016, the alarming shortages in Cape Town in 2018, and the West's unprecedented 2020 fire season should remind us that the difference between plenty and shortage is often only a few dry seasons away. As so many water managers, scientists, journalists, and concerned citizens have suggested, current practices must change. We must invest in new and better technology to reduce and save water. We must be creative and accept stark realities in how we adopt policies to ensure a sustainable future. Eric Kuhn and John Fleck put it this way in their recent treatment of the failure of the early framers of the 1922 Colorado Compact to adhere to data witnessing a history of drought and low flows throughout the watershed: "Innovation, cooperation, and an expanded reliance on science are now the foundation for basin-wide solutions" (217). Indeed, a greater willingness to accept the fact that the river likely has only thirteen million acre-feet (maf) per year, far below the estimates that have governed river management in the past, is essential if we are to be successful moving forward (213). Indeed, as *Science*

Be Dammed (2019) demonstrates, the failure to establish and apply sound scientific findings within Colorado River management in favor of politics and boosterism has led to many of the resource challenges we face today as we work with an overallocated river.[7]

However, the innovation and adaptation to enact the needed shifts in water use are quite difficult to implement without a broad reorientation at the most fundamental level about what water means to each of us. What value does water have for us individually? As a culture? While science is essential in shaping policy and providing the clearest view of what current conditions are like on the ground, it is less adept at motivating the general populace to alter behavior. Consider the precariousness of our freshwater resources. For many of us, we are so accustomed to clean water coming out of the tap that it is all too easy to ignore ample evidence that highlights how freshwater systems across the globe are in peril. The reliability of this luxury renders the realities nearly invisible, and no amount of "truth telling" seems to make a difference. Thus, while science performs a crucial role in identifying and making sense of current environmental challenges, it cannot do this work alone. And it certainly can't do it alone in a society where so many discount scientific findings as just another example of "fake news." Therefore, while we need the best science possible along with new laws, technologies, and economic incentives to address current and future water-management issues within the Colorado River Basin, we also need a monumental attitudinal shift in how we envision water. Ultimately, an alternative water ethic to the status quo must emerge. We need to think and talk about water in ways that capture the true importance of this limited resource. Indeed, how we imagine and communicate about water is essential to this reorientation, for as ecocritic George Handley astutely observes, "Perception [is] the last great clean and renewable resource" (*Home Waters* 212). This study seeks to initiate a shift in perception away from the historical attitudes and their respective practices that have governed Colorado River management by attending to the values about and relationships with the river expressed by traditionally marginalized perspectives from the basin. These voices, and their reliance on narrative or story, offer alternative means for understanding the river and the present challenges it faces today that science, policy, economics, and law fail to capture.

Consider the following historical example about naming a river that illustrates just how powerful language can be in altering our reality. Since the arrival of Anglo-Americans to the region, the Colorado River referred to the river below the confluence of the Green and Grand Rivers in southeastern Utah (Hundley, *Water* 141).[8] As Russell Martin explains in his history of Glen Canyon Dam, the section of river that flows through the state of Colorado today (previously known as the Grand River) has widely been accepted as the river's headwaters, despite being a few hundred miles shorter than the Green River flowing out of Wyoming and through eastern Utah (34). A quick perusal of maps, government publications, news services, and websites about the river corroborates this point.[9] Another Colorado River historian, Philip Fradkin, takes up this controversy citing E. C. La Rue, the head of the US Geological Survey who, in his 1916 report *The Colorado and Its Utilization*, noted that the "Green River drains a larger area than the Grand and is considered the upper continuation of the Colorado" (qtd. in Fradkin 35). If we assume that the USGS is the arbiter of all things riverine, then it follows that because the "longer stem of a river, termed the master stem, is usually considered its main branch," the Green should accordingly be the master or mainstem, while what is known as the Colorado today is its primary tributary (36). However, five years following La Rue's report, some rhetorical manipulation magically shifted the Colorado River's headwaters hundreds of miles to the southeast and into the state of Colorado.

In the early 1920s, the seven Colorado Basin states postured to get what they saw as their fair share of the river's water. As each state developed a plan for how best to allocate the river, Colorado, Utah, and Wyoming considered renaming their respective tributaries to give the impression that they were indeed the headwaters state, the home of the Colorado itself. As Russell Martin observes, "It seemed logical to members of the Utah legislature that a political as well as a rhetorical advantage could be pocketed if the Green were renamed the Colorado, the true river then officially bisecting their beloved state" (34). While Wyoming negotiated with Utah to facilitate this change, Colorado was quicker to act. Congressman Edward Taylor lamented that the river's current appellation as Grand was a "most unjust, unnecessary, and unfortunate misnomer and absurd incongruity" (United States 8–9). Through a lengthy outpouring of state pride, Taylor expounded the state's many geographic splendors, concluding that this most significant river deserved to be known for the state boasting such an embarrassment of riches. Compelled by his effusive rhetoric and extensive documentation witnessing to the state's clamor for this renaming, Congress ratified the change, which Warren G. Harding signed into law on July 25, 1921 (Hundley, *Water* 141; Fradkin 35; United States 8–9, 28–32).[10] Thus, through political jockeying and rhetorical deftness, the shorter of the two stretches received the nation's blessing as the true headwaters of the Colorado, while the Green became the principal tributary. The impacts of this rhetorical manipulation were significant. Even as the state of Colorado provided the lion's share of the river's flow—a point well known to the water managers—the renaming cemented within the public's mind that Colorado is *the* headwaters state and therefore the home of the Southwest's most important water source. This recognition certainly did not hurt Colorado's allocation requests during the 1948 Upper Colorado River Compact negotiations, which divided the waters among the states.[11]

This little-known event powerfully exemplifies how language and its manipulation can have significant material impacts upon a river and our relationship to it. Key to the river's ongoing interaction with humans is the variety of often competing narratives that have dictated how the river would be named, imagined, and used by the American public in literature, journalism, public policy, and science. Like the river itself whose many small creeks and streams mix and mingle to produce the mainstem, these different perspectives combine to carve out a unique body of representations about the river and its surrounding landscapes. However, just as some tributaries (the Grand/Colorado) have historically taken precedence over others (the Green), so too have certain versions of these perspectives—at the expense of others—dominated which uses and representations of the river are most "appropriate" for the nation. Thus, a distinct hierarchy of voices has developed since the nineteenth century as a result of controversies over human engagement with the river, pitting different understandings of what and whom the river is for against one another. One of the primary goals of this study is to reclaim some of these marginalized perspectives and consider how they engage with and often reject the dominant visions of the Colorado River and the larger American West as a region of limitless plenty.

Such contrasting needs, caused in large part because the Colorado River Basin comprises some of the most arid lands in North America, have made the river a magnet for controversy. As Philip Fradkin notes in *A River No More* (1968), "The Colorado is the most used, the most dramatic, and the most highly litigated and politicized river in this country, if not the world" (15). Writing in his classic on western water development, *Cadillac Desert* (1986), Marc Reisner corroborates Fradkin's view, stating, "The Colorado's modern notoriety... stems not from its wild rapids and plunging canyons but from the fact that it is the most legislated, most debated, and most litigated river in the entire world." As if these accolades aren't enough, he goes on to point out that "[the Colorado] also has more people, more industry, and a more significant economy dependent on it than any comparable river in the world" (120). The result of the countless competing claims placed on the Colorado since Anglo arrival to the American Southwest is a body of textual production that is unsurpassed in the annals of river literature. "You could create a second river out of all the ink that has been used to write about every aspect, every nook and cranny, of this once-great river," as Charles Bergman explains in his own treatment of the river's terminus (39). Thus, not only is the Colorado a material entity flowing through and shaping a region of inestimable geologic variety, but it is also a river of words, a stream of signs coursing through a complex imaginative landscape that continues to evolve as it encounters shifting political and social terrains. Over time, this secondary imaginative river has come to carry the hopes and dreams of a region while also articulating the potential demise of a society accustomed to excess.

From the competing interests that have long dictated the Colorado's use emerges a "crisis of meaning," what T. S. McMillin describes in *The Meaning of Rivers: Flow and Reflection in American Literature* (2010) as a problem associated with how we think and act toward rivers. Through his examination of what literary representations of rivers suggest about how we make sense of the world (xii), McMillin explains, "As a culture, because of the diversity of groups that constitute that culture and because of the knottiness of the subject, we have not collectively made up our minds about what rivers mean—which implies that the possibility exists of changing the way we think about rivers." Our inability to reconcile the apparent disparities in how our culture views rivers has led to what McMillin sees as a "crisis in meaning to go along with serious environmental crises." Unable to settle upon what rivers mean to our culture—are they conveyors of spirituality, sites of recreation, entities for exploitation?—we find ourselves pulled by competing narratives, many of which contradict one another and forward narratives and ideologies that fail to cultivate a more sustainable ethic about water and our nation's rivers. Thus,

McMillin argues that the stories we tell within our national literature are a cause for the environmental crises we currently face, that "those crises [of meaning and the environment] are related to one another" (xviii). If we accept this premise to be true, how we think, write, and communicate our ideas about rivers has a bearing on the actions of those engaging these representations. As the following discussion suggests, nowhere is such a crisis of meaning, or as I argue here a crisis of representation, more evident than within the textual production of the Colorado River.

Riding the Waves of Myth with the "Lords of Yesterday"

Ironically, while such crises exist throughout the Colorado River watershed, they don't exist entirely because of a lack of cohesive meaning about the river, as McMillin generally suggests. While it is true that this particular river has been and continues to be represented by countless entities whose purposes are often at odds with one another, a rather unified understanding and approach to the river's significance has persisted over the past two centuries. In fact, as I argue in this book, the environmental crises affecting the Colorado appear less from a lack of shared understanding than a lack of vision regarding the broad spectrum of meaning that this river embodies. Ultimately, it is the relatively static representations of the river that dominated the literary and discursive constructions of the river in the late nineteenth and early twentieth centuries that continue to limit a broader, more inclusive awareness of the river's many meanings today.

Since the nineteenth century, the frontier, garden, and wilderness myths have powerfully informed the textual production of the Colorado River. They are "stories drawn from a society's history that have acquired through persistent usage the power of symbolizing that society's ideology and of dramatizing its moral consciousness" (Slotkin 5). In the case of the Colorado, these myths have influenced a long line of explorers, scientists, engineers, boosters, politicians, and literary writers whose own texts have further perpetuated these myths, thereby reinforcing them within our culture and influencing how the river's many stakeholders use its water. This transmission from thought to practice is embodied in what legal scholar Charles Wilkinson calls the "lords of yesterday, a battery of nineteenth-century laws, policies, and ideas that arose under wholly different social and economic conditions but that remain in effect due to inertia, powerful lobbying forces, and lack of public awareness" (*Crossing the Next Meridian* 17). One of these relics of the past that carries immense weight for western water management is the prior-appropriation doctrine, the "first in time, first in right" principle built on the premise that "the first to use water from a specific source held a prior right that would be protected against the claims of others" (Getches 83). This policy, Wilkinson claims, is "perhaps the area in which the law of the West is most out of kilter," as it ensures that senior rights holders can appropriate or use their entire share of rightfully claimed water even if that means that those who hold junior rights live upriver from the senior holder (*Crossing the Next Meridian* 21).[12] For the junior rights holder, this means they must watch the water rush by and perhaps even see it disappear altogether if there is only enough water to meet the prior appropriator's share.

The inertia to which Wilkinson refers relies upon the circulation of myths and the various modes of discourse (whether written, oral, or in other means of communication) that disseminate them. The most iconic of these myths is the West as frontier, a limitless place of adventure and opportunity. Most famously articulated by Frederick Jackson Turner in his address to the American Historical Association at the 1893 Chicago World's Fair, the frontier myth had influenced American thinkers and writers such as Crèvecoeur, Franklin, Emerson, Lincoln, Whitman, and "a hundred others" long before Turner's speech (H. Smith 3). Like his predecessors, Turner looked to the nation's expansive, open land as the site wherein "this perennial rebirth, this fluidity of American life, this expansion westward with its new opportunities, its continuous touch with the simplicity of primitive society, furnish the forces dominating American character" (F. Turner 3). For him, the frontier, "in spite of environment,... did indeed furnish a new field of opportunity, a gate of escape from the bondage of the past; and freshness, and confidence" (38). Steeped in the optimism of the nation's westward migration and settlement, these observations further solidified within the national imaginary the belief that no matter the challenges the West's geography, climate, or Native inhabitants may present, endless opportunity and bounty awaited.

As settlers entered the arid West, they found the lack of water to be one of the most immediate hurdles in their quest to secure their dreams of re-creating this wasteland into a fertile oasis. Contrary to the image of the West as the Great American Desert, the garden myth "took the form of a proliferation of notions about an increase of rainfall on the plains" (H. Smith 179). A common mantra for nineteenth-century boosters and settlers in the arid West was that "rain would follow the plow"—a belief suggesting that Providence would further the nation's destiny of civilizing the continent by bringing increased rainfall to the region. A closely related hope emerged in the promise of reclamation and what irrigation could bring to the region's parched lands. Many advocates looked to the arid West with the words of the Old Testament prophet foremost in their imaginations of what the region could be: They wanted to make the desert "blossom as the rose" (Isa. 35:1).

This biblical prophecy aptly describes William E. Smythe's *The Conquest of Arid America* (1899). As one of the great proponents of reclamation, Smythe dedicates his work to "those who have the courage of their optimism—for the homeseekers who...are to grapple with the desert, translate its gray barrenness into green fields and gardens, banish its silence with the laughter of children" (*The Conquest* x–xi). Tapping into the optimism that undoubtedly infuses Turner's frontier thesis penned only a few years prior to *The Conquest*, Smythe views the West's reclamation as the fulfillment of a divine partnership between God and man to further civilization's and nature's progress. He writes, "The man who works intelligently in creating his irrigated farm with the raw materials of land and water, knows that in this smaller sphere he is engaged in finishing the world" (329). By bringing the life-giving waters to the West's parched lands, Smythe saw America's reclamation investments as a continuation of divine creation.

For those laboring to redeem the West's deserts, reclamation was more than just a transformation of the land; it was part of a conversion that would lead to a more enlightened society. As Smythe further notes of the grand designs for the West, "It was the destiny of the [West] to lie fallow until humanity should feel a nobler impulse; then to nurse, in the shadow of its everlasting mountains and the warmth of its unfailing sunshine, new dreams of liberty and equality for men" (*The Conquest* 19–20). However, such dreams were not possible for many American citizens. Writing for the Reclamation Service in 1918, C. J. Blanchard echoes Smythe's vision for the American West as he promotes the region's irrigation: "To the new empire in the West have flocked the young, the strong, the adventurous, and herein we are witnessing a gradual welding of all the Aryan races into a final race type. Signs are not lacking that this type in time will dominate the world, for the desert offers to every man his true birthright—room to breathe, sunshine, a sure reward for intelligent labor, the individual home, and an opportunity to become independent" (4).

Today, we shrink at this rhetoric. Yet this view dominated the late nineteenth and early twentieth centuries as the nation looked to establish itself as a world power. In fact, a closer look at reclamation history in the arid West suggests that such beliefs and their associated exclusionary practices exist even today, evident in the struggles by Native Americans to assert their water rights, for example. Thus, while Pike's Great American Desert once rebuffed settlers, the "miracle of irrigation" made it possible for man to facilitate the region's transformation into what Mark Fiege dubs an "irrigated Eden" (41–47). For Fiege, "the garden myth became an epic of personal and national regeneration.... Triumphing over chaotic wildness, re-creating the lost Eden, the pioneers redeemed themselves and the land, restored the agricultural base of the Republic, and realized God's plan for the earth" (171). Striving to return to the garden once lost through the Fall, nineteenth-century beliefs about the arid West dictated man's divine right to drastically alter the region for the onward march of progress and civilization.

Like the myth of the garden, the wilderness myth in American culture also has connections with the frontier. In many ways, we may consider the wilderness myth as an extreme outgrowth of the frontier's emphasis on individualism and the renewal that comes from entering uncivilized land. Similarly, we can read it as a critical response to the development and "progress" that the frontier myth brought to the nation's lands. In *Wilderness and the American Mind* (1968), Roderick Frazier Nash explains that where wilderness once was something that Americans sought to conquer as it "constituted a formidable threat to... survival" and tempted one to "[revert] to savagery," rapid settlement and cultivation of the nation's lands would change these attitudes (24). While Nash grounds a wilderness ethic in Romanticism, the nineteenth century's pervasive industrialization and Turner's call for the closing of the frontier in 1890 further entrenched the notion of wilderness as myth.[13] As "too much civilization, not too little, seemed at the root of the nation's difficulties," wilderness became a symbol for hope, a place in which to distance oneself from society's ills (143). It too became a "source of virility, toughness, and savagery" and a site endowed with "aesthetic and ethical values," which provided "opportunity... for contemplation and worship" (145). As the nation moved into the twentieth century and embraced a growing environmental ethos, the wilderness myth became ever more entrenched in the American imagination as it positioned wild nature as the untainted embodiment of what is true and real, a place to escape the facade of the modern life. As such, the wilderness myth has taken on different guises in its relation to the West. It first depicted the region's open lands in need of cultivation and civilization as its wild recesses were something to fear. Once the garden myth and other factors helped tame some of these lands, wilderness then became something to celebrate and preserve.

Tributary Voices

Over the past century and a half, these historical perspectives have reflected an imbalance of representation that has privileged a narrow range of interests, genres, and voices while disregarding those that fall outside the dominant vision expressed by these myths. The results of this activity speak to McMillin's crisis of representation and meaning, leaving us to flounder in a morass of competing narratives. Similar reactions to the region's current views about water have extended beyond the literary realm to law. At a recent symposium on western water law, legal scholar Tom Romero suggested that when we talk about issues of diversity in regards to the arid West and its resources, "we don't necessarily have a common language or a common framework" ("Welcome Address"). For far too long, westerners have argued and fought over water and the "first in time, first in right" dictum.

Thus, one of the primary goals of this study is to respond to Romero's observation by offering an alternative framework, one built upon a broader, more egalitarian vision of the river that relies on the power of story or narrative as they intermingle with scientific, legal, and procedural discourse to create a new reality and vision of living with the Colorado. To do this, I argue that we need to put reclamation to use in ways that urge us to grapple with the material implications brought about by the 1902 Reclamation Act and by its advocates such as Smythe and Blanchard. Rather than pursuing a continued redemption of the land through the application of water, we must embrace a discursive reclamation, one that seeks to reinvigorate the imagination and the texts we use to convey those visions in hopes that a more equitable, sustainable, and even restorative relationship with the river might take root and flourish. In essence, I advocate for a discourse ecology of water literacy.[14]

This discursive ecology invites a broader assessment of river representations from across many different disciplines and forms of expression to enhance our understanding of Colorado River issues. In essence, a discourse ecology of water literacy relies upon a wide range of genres as well as the arguments within those genres to reorient our thinking about and potentially our actions regarding the river. This methodological approach draws from recent scholarship in the environmental humanities (to be addressed in greater detail in chapter 1) that seeks to "foreground how interdisciplinary discussions of narrative, visual, and creative works from different regions can advance understanding of the specificity of ecological concerns as well as anticipatory visions for the future" (DeLoughrey, Didur, and Carrigan 2). Such identifications and reimaginations are central to this discursive ecology of water literacy and manifest through what I call "tributary voices" or the neglected and lesser-known perspectives of the river that have been traditionally relegated to a figurative backwater—a space of critical silences whose perceived absences nonetheless denote their presence.

My conceptualization of tributary voices draws from a number of perspectives, some of which I touch upon here and others developed also in chapter 1. Tributary voices embody various hydromythologies, what literary scholar Rebecca Lynne Fullan describes as "the way core cultural beliefs about and understandings of water create stories, relationships, and actions" (137). Each tributary voice considered in this study forwards a particular orientation in its respective consideration of the river and the environment, and I seek to examine these considerations for their relevance to informing present approaches to water use. We can also think of tributary voices as unique discourse communities or "environmental networks" that constitute, as policy experts Raul Lejano, Mrill Ingram, and Helen Ingram write, "how people understand relationships that bind them with each other, and with animals, resources, and other elements of the environment" (1). Central to the formation of these relationships is story, which, in the case of "environmental narratives," offers us "important places to look for practical inspiration about new ways of living with the world" (3). Such inspiration often comes as these networks challenge the status quo while they work within expected norms to promote alternative water ethics and action. In this light, tributary voices operate as heterotopias, which Catrin Gersdorf defines as "spaces situated *outside the center* of social and political order at a given historical moment *yet not totally disconnected from that center*. That is, the reformatory (or even revolutionary) impulses emanating from heterotopic spaces are as much an effect of *resistance to* as they are *dependent on* contact with the spaces representing dominant forms of social and cultural ordering" (237). Like a river's tributaries that mix and mingle, often in turbulent ways with the dominant (mainstem) reaches of a water course, these tributary voices converse with and challenge genres that have traditionally shaped western water use and management emblematic of the sciences, law, economics, and policy.

At the same time, these tributary voices in varying degrees engage the discourses reflective of settler-colonialism, which has harnessed such genres to promulgate the visions expressed by the pervasive western myths previously noted. Tributary voices are both dependent upon the dominant river discourses and resistant to them as they negotiate the river's and region's myths, the disciplines governing water management, and even the genres used to convey these perspectives, whether they be exploration narratives, fiction, histories, procedural documents, or scientific reports. As they engage these dominant perspectives and discursive modes, the tributary voices advocate ways of understanding and interacting with the river that are not recent phenomena. In her "Grounds of Hope," Rebecca Solnit challenges the belief that "the solutions we need have not yet been launched or invented, as though we are starting from scratch" when it comes to addressing the social, political, and environmental crises of our day. Rather, as the tributary voices often demonstrate, the means by which to understand and interact with the river have existed all along. "The real goal," Solnit explains, "is to amplify the power and reach of existing alternatives. What we dream of is already present in the world" (34). This is the case with a number of the tributary voices examined in later chapters as they reveal alternative epistemologies and practices that have run concurrent with and at times long predate the lords of yesterday and that offer pathways by which to both acknowledge the challenges before us today and offer hope for the future.

The Power of Story

Central to creating a discursive ecology of water literacy of the Colorado River Basin regardless of whether we speak of tributary voices, hydromythologies, environmental networks, or even heterotopias is the role that narrative or story plays in shaping one's relationship to the river. In their introduction to a collection celebrating and seeking to protect Utah's wild landscapes, Stephen Trimble and Terry Tempest Williams place their trust in "the power of story to bypass political rhetoric and pierce the heart" (*Testimony* 7). Williams elaborates on the role that story plays in shaping perception in a later book about the Colorado Plateau, the uplifted landscapes that constitute much of the Four Corners region and the heart of the Colorado River Basin. She writes, "Story offers a wash of images and emotion that returns us to our highest and deepest selves, where we remember what it means to be human, living in a place with our neighbors" (*Red* 3). While stories—evident in imaginative literature, most obviously, but in other types of communication as well—are compelling, they do not typically play a significant role in resource management and governance. Whereas scientific and legal discourses most often exemplify—at least for the dominant culture—the means through which fact and truth are established and quantified, story is viewed as too subjective, too emotional, to play a role in making serious decisions. Thus, access to "the way things are" in regards to our water bodies has prevented groups like American Indians whose access to such discourses has been historically limited and whose primary framework for constructing their reality—storytelling via oral tradition—has been disregarded as myth.[15] In sum, stories are an essential part of a broader discourse ecology that seeks to engage many different perspectives in order to form a more comprehensive reality for the myriad ways in which the river is valued and used.

However, as the following chapters illustrate, stories about water and one's relationship to it as expressed through a narrative aesthetic offer us a unique and needed perspective when it comes to managing a threatened river. Indeed, as Elizabeth DeLoughery, Jill Didur, and Anthony Carrigan argue in their collection on postcolonial ecocriticism, "Attention to narrative and aesthetic form is fundamental to understanding environmental crises," because "understandings of the environment are embedded in language, narrative, history, and the cultural imagination" (13). As human problems, environmental crises need answers that derive from the same source creating the problems—through perceptions, beliefs, values, and so forth—causing the crises in the first place. Narrative's value derives from its ability to invite those receiving the story to pause and reflect, identify with shared experiences, offer alternative solutions and realities to the current situation, and ask difficult questions for individual and collective contemplation. While the formative discourses of law, economics, science, and policy and their associated set of regulations, cost-benefit analyses, reports, and models are essential to sound governance, how effective are they at inviting the general public to have a change of heart about water use? How do longtime westerners and the millions who have moved to communities reliant on the Colorado River from wetter climes "get over the color green," as Wallace Stegner famously wrote, to embrace a lifestyle more accepting of aridity, which he identified as "the primary unity of the West" ("Thoughts" 54, 47)? Or, as April Summitt has asked in her recent environmental history of the Colorado River, "Why do people press on with unsustainable policies in the face of impending disaster? How does a population change its attitudes, perceptions, and laws relating to a scarce resource before catastrophe strikes?" (xii). Because our water resource challenges are so complex, no one solution exists. But as *Tributary Voices* argues, the reclamation of the river's marginalized perspectives and their various stories becomes a powerful tool in initiating the changes Stegner, Summitt, and so many others seek.

Indeed, narrative is an essential element in rethinking our connections to the places we inhabit and the resources upon which we depend. Here I draw from Italian ecocritic Serenella Iovino and her own reflections on moving to Italy's Po River valley, a region that, like the Colorado, has experienced significant environmental degradation as endless demands have compromised the country's most important waterway. She reminds us, as McMillin has above, that environmental ruin is symptomatic of larger cultural issues: "The ecological crisis of place is not limited therefore to its being 'disrupted and injured' in its organic balances. It is a crisis that involves a place's ecology of mind, namely, its imagination" (105). To mediate these concerns, Iovino advocates for a "narrative reinhabitation," which she defines as "a cultural-educational practice that consists of restoring the ecological imagination of place by working with place-based stories." *Tributary Voices* demonstrates the value of reinhabiting the Colorado River Basin through narrative to restore our collective ecological imagination of this most important region. As Iovino further explains, "Visualizing the ecological connection of people and place through place-based stories is a way to remember a dismembered unity, to enliven our cultural and ecological potentialities—to reanimate the world" (106). By sharing the contributions that the selected tributary voices make to our understanding of the Colorado River and surrounding landscapes, we broaden our horizons as to the range of possibilities that await us as we better integrate the needs of the watershed's many constituencies.

To demonstrate what constitutes a discursive reclamation or a narrative reinhabitation of the Colorado River Basin, I turn to recent conversations in the environmental humanities and their contribution to addressing current environmental crises. Like Summitt and many other humanities scholars who are bringing greater interdisciplinarity to their respective fields, I seek to shed increased light on the Colorado's discursive complexity by forging connections between literary and rhetorical studies, history, geography, science, law, and policy to more thoroughly examine the contributions of the river's marginalized perspectives and texts. My methodology here is highly informed by the work of Helen Ingram, who, more than twenty years ago, proposed a "humanists at the headgates" approach to addressing western water issues. She explains:

> At the same time that nations, states, regions, and cultures have battled over water, another kind of warfare has been going on. Different professions, disciplines, and perspectives have vied with one another to become the dominant model for framing water issues and their solutions.... It is my contention that the inadequacy of these modes of thinking is responsible for the serious water problems we now must solve in an atmosphere of suspicion and hostility. The engineers, the lawmakers, lawyers, and political power-wielders, and the economists have spoken with conflicting voices, different images, and very different prescriptions about what we should do concerning water problems. In the meantime, the voice of the humanists speaking through literature has been all but ignored. (157–58)

While significant gains have been made along the Colorado to invite a greater diversity of perspectives over the past two decades, more can be done to increase the participation of marginalized voices and advance a broader, more holistic approach to water decision making. Through her extensive analysis of western water conflicts, Ingram further contends, "When engineers, politicians, or economists have taken control of water-resources decision making, a singular, exclusivist perspective has dominated. Humanists, it seems to me, would be better gatekeepers because they are respectful of value pluralism.... [H]umanists would emphasize the rights of ordinary people to participate in the public deliberations that should govern water choices" (166).

In sum, *Tributary Voices* seeks to reenergize Ingram's call to action, as we need a "humanists at the headgates" view of water resource management now more than ever. If we are serious about understanding how our written expressions about the river have shaped and can influence the watershed and bordering regions physically and ideologically, we must consider a broader range of discourse beyond those that have historically dictated river management and allocation. Likewise, we need the contributions of disciplines such as those in the humanities that water resource circles often overlook but that have significant power in helping us contemplate values, reconsider behaviors, and work toward a more egalitarian use of the river. We need the added perspectives of humanists to consider how various discourses of dominant and marginalized voices—as environmental networks—influence how the public views (or fails to view) the river that will determine the basin's destiny. Such contributions by these fields and the narratives they engage will also help us build toward a middle ground that "allows us to move away from grand narratives of human technological transcendence and conquest on the one hand, and strict preservation and wilderness on the other" (Lejano, Ingram, and Ingram 174). With a clearer understanding of the dominant ideologies that have shaped the river and a commitment to the contributions of tributary voices, we find ourselves in a better position to sit down at the bargaining table as equal stakeholders and address the myriad challenges that will dictate our relationship and reality with this most unique river and region. As we approach the one hundredth anniversary of the signing of the Colorado River Compact, it is time to reassess the past and consider what stories both old and new can guide the river's future.

While there exist far too many tributary voices past and present to address in this study, those examined in subsequent chapters demonstrate the potential of a discourse ecology of water literacy to reshape present understandings about the Colorado River and identify various ethical frameworks by which to approach future river management and use. Aside from chapter 1, the book follows an organizational pattern that loosely mirrors trends in ecocriticism over the past three decades, as it begins with the watershed's earliest written accounts and the nature-writing tradition to move toward more diverse perspectives in terms of gender, race, and faith. However, unlike a river journey that moves linearly from the point of entry to the final takeout, the book's chapters do not have to be read sequentially to see the value of tributary voices. I invite readers to choose those voices that are of most interest, to linger where desired, and put-in elsewhere when prompted.

In chapter 1, I outline the theoretical foundation for this project and articulate the need to expand beyond the typical representations that have traditionally portrayed the Colorado River Basin. Drawing on recent scholarship in the environmental humanities over the past ten years or so, and employing its interdisciplinary approach to environmental critique, I situate this discursive ecology within New Western History, critical regionalism, bioregionalism, and the pragmatic rhetorical tradition to bring together fields such as history, geography, law, literature, and rhetoric. Within this confluence of disciplines, I transpose the idea of a watershed and its various definitions onto Walter Beale's grammar of motives to suggest how the watershed operates as both a metaphor and a physical entity that implies and delineates boundaries. Mapped onto a rhetorical framework of how language shapes our perceptions and realities, the watershed becomes a powerful symbol by which to identify, account for, and analyze the multiple cultural, gendered, national, racial, and social perspectives about the Colorado River. This convergence of ecological and rhetorical frameworks not only deepens our appreciation for the diversity of voices from the Colorado River Basin but also emphasizes the value that rhetorical studies has in the growth of ecocriticism and the environmental humanities more broadly.

Chapter 2 begins this reassessment of Colorado River knowledge through the basin's tradition of science writing emanating from the watershed's deserts that spans a period from Spanish exploration to contemporary nature writing to include accounts by Mary Austin, Ann Zwinger, Joseph Wood Krutch, and Edward Abbey. I locate Craig Childs's *The Secret Knowledge of Water* (2000) within this genealogy and examine how his use of paradox creates a niche within the desert-writing tradition that allows the desert and its water to exert agency as he negotiates competing ways of knowing the world. I claim that his use of paradox creates a physical and philosophical contact zone, one where the reader confronts the borders that mark systems of knowing, divisions between the human and the nonhuman, and the strategies and positions used to evoke the mystery and wonder of the watershed's deserts. Reading *The Secret Knowledge of Water* through these lenses encourages us to rethink our relationships with aridity and the vast spectrum of matter (human and nonhuman alike) shaped by this reality with the broader goal of identifying how we might more effectively imagine and create a more sustainable relationship with our surroundings.

Chapter 3 continues the theme of exploration but juxtaposes the male-dominated genre of Colorado boating narratives made famous by John Wesley Powell's landmark 1895 account and replicated well over a century later by many other men with a growing body of women's river writing. Specifically, I attend to texts by Patricia McCairen, Laurie Buyer, Louise Teal, and Kathleen Jo Ryan and their journeys on the Colorado through the Grand Canyon to showcase the many women exploring and writing about the river. Through their unique use of story and various genres, including memoir, fiction, interviews, and poetry, these women appropriate and critique the male-dominated western to create a space for the female river experience. In doing so, they urge readers to embrace a vision of the New West, one that challenges the region's legacy of nostalgia and conquest to offer an alternative narrative grounded in inclusivity, humility, and restraint.

Chapters 4 and 5 coalesce around the theme of environmental justice along the Colorado River evident in the efforts by Hispano, Latina/o, Mexican, and American Indian stakeholders to express their connections to the river and to become more visible participants in river governance. The impact of Navajo Dam's construction on northern New Mexico's Hispano communities in the 1960s and the ways in which the broader Latina/o community on both sides of the US-Mexico border has worked to define its relationship to the Colorado River Basin in recent years define chapter 4. Central to this discussion are the concepts of *buen vivir* and *querencia*—the appreciation of living well and the cultivation of a sense of place—that critique notions of pristine wilderness that have long influenced the region's anti-dam movement to create a space for locally based conservation efforts. This chapter asks us to consider what we might gain by looking at Navajo Dam and its creation as a more appropriate symbol for the environmental movement than the oft-celebrated and -demonized Glen Canyon Dam. It also attends to the oft-neglected Mexican perspective and its role in the river's management through the founding documents of the Asociación Ecológica de Usuarios del Río Hardy y Colorado (AEURHYC), or the Hardy and Colorado Rivers Ecological Water Users Association, a local water users' association within the Colorado Delta, working to protect and restore the delta's fragile ecosystem. Through these two examples, the chapter invites readers to consider what it means to live the "good life" and how recognizing their *querencia* might alter how they imagine their relationship to the basin and all its inhabitants up and down the river.

While Navajo Dam displaced Hispano communities, it also enabled the Navajo Nation to develop its long-standing water rights to achieve greater economic security and prosperity. Chapter 5 addresses the Navajo's dam campaign as an introduction to a larger conversation about the value of Indigenizing environmental justice as it relates to the basin's Indigenous tribes' and nations' efforts to assert their legal water allocations. Examples in this chapter include novels and poetry by Leslie Marmon Silko, Ofelia Zepeda, and Natalie Diaz; a water study by the Colorado River Basin's Ten Tribes Partnership; and the recent "Voices of Grand Canyon" project by the nonprofit Grand Canyon Trust. This wide array of Indigenous voices and texts demonstrates the various methods by which Indigenous communities, tribes, and nations are working through the injustices of the past to become significant players in today's water landscape, thereby giving new meaning to the "first in time, first in right" principle as they assert their rhetorical and water sovereignty.

Building on the discourses of *buen vivir*, *querencia*, community, and stewardship in previous chapters, chapter 6 makes a case for the role religion can play in discussions of sustainability. It examines the influence of the Church of Jesus Christ of Latter-day Saints' water development in the Colorado River Basin; the contributions of John A. Widtsoe, who, in the first half of the twentieth century, was not only a prominent leader of the church but a world-renowned chemist and expert on crop production in arid regions; and the church's promotion of provident living and earth stewardship. As a major player in western reclamation and the transformation of the Colorado Basin historically, the church may seem an unlikely partner in stewarding a more responsible water ethic in the region. However, by examining the church's early water practices and leadership within the nation's reclamation efforts in light of its theological perspectives, we see how religious belief and practice offer another important set of resources from which to build a more sustainable future within the watershed.

The epilogue concludes this study with a look at the art of Allen Tupper True and poetry of Thomas Hornsby Ferril found in the Colorado Capitol rotunda to reimagine the Colorado River's future through the contributions of the tributary voices examined herein. These perspectives and the various genres, which articulate their connections to the river, enable us to reenvision the river and watershed as a more complex discursive space, one that offers us alternative ways and means of talking about and interacting with these entities. Moreover, with such heightened awareness of these values, there is the hope that we can cultivate a more unified and ethical approach to water management that will sustain all the basin's inhabitants for generations to come.

CHAPTER ONE

"Humanists at the Headgates"

Mapping a Discourse Ecology of Water Literacy

To know the spirit of a place is to realize that you are a part of a part and that the whole is made of parts, each of which is whole.
—Gary Snyder, *The Practice of the Wild*

Before launching our proverbial boat into the turbulent waters of Colorado River discourse, I wish to outline in greater detail the frameworks and theoretical positions that illuminate the manifold ways the tributary voices considered herein employ narrative to establish their relationships with the Colorado River and surrounding landscapes both historically and in recent years. As Ingram suggests in the previous chapter, current water-resource challenges demand an "all hands on deck" approach—no one discipline or limited range of perspectives should dominate how the intellectual community and the greater public understand water issues. Thus, this chapter establishes the various ideas that constitute a "humanist at the headgates" vision for the Colorado River. Central to this discussion are the environmental humanities, a coalition of scholarly and activist approaches that are currently shaping numerous environmentally focused fields. Those especially relevant to establishing a discourse ecology of water literacy include perspectives from western American history and literature, environmental literary studies/ecocriticism, and rhetorical studies. Together, they illuminate useful heuristics such as "New Western Rhetoric," the watershed model, and Walter Beale's semiotic grammar of motives by which to challenge some of the most pervasive and dominant historical myths, discourses, and genres that have dictated how the river has been imagined over time. In so doing, this confluence of perspectives loosely tied together under the rubric of the environmental humanities helps us identify tributary voices and assess the value they and their respective narratives offer, and it invites us to critically listen to the potential solutions they provide. Through this process we can begin to reimagine and redefine the contours of our own relationships to the river and watershed both individually and collectively in a manner that makes us more aware of and invested in the future health of this remarkable water body and region.

The Environmental Humanities

In light of the ongoing drought and many other challenges facing the Colorado River today, many believe that more science and better technology will lead us out of our current dilemmas rather than looking to disciplines within the humanities for possible answers. In their consideration of the environmental humanities, a panel of international scholars acknowledges this common assumption: "When policy makers and mainstream media outlets seek expertise on the environmental crisis today, they seldom turn to environmental historians and philosophers, much less to ecocritics" (Bergthaller et al. 262). Indeed, a 2019 article by Steven Allison and Tyrus Miller, an ecologist and dean of humanities, respectively, acknowledges this bias, noting how the recent Intergovernmental Panel on Climate Change report leaves little room for how ethics and social justice might play into the scientific and technological solutions outlined within the document ("Why Science Needs the Humanities"). These observations should be of little surprise, as western society emphasizes objectivity and rationality, the supposed terrain of the sciences, over the humanities' attention to subjectivity and emotion to provide the necessary answers we need to address society's environmental failings. Make no mistake. We need the best scientific thinking possible. However, while more science and better technology are essential to future efforts to more effectively govern the river, they are certainly not and cannot be the only solution.

One of the primary reasons for a more comprehensive approach is that while science, law, and policy excel at explaining what is happening to our world and establishing rules to govern action, they are less equipped to address human responses and the beliefs and values that influence such conclusions. Numerous scholars working on environmental solutions have noted this need for a more inclusive, interdisciplinary approach to address the challenges of the Anthropocene, our new human-dominated era. In a 2012 issue of *BioScience*, Sverker Sörlin explains that "in a world where cultural values, political and religious ideas, and deep-seated human behaviors still rule the way people lead their lives, produce, and consume, the idea of *environmentally relevant knowledge* must change. We cannot dream of sustainability unless we start to pay more attention to the human agents of the planetary pressure that environmental experts are masters at measuring but that they seem unable to prevent" (786).

Others have corroborated this view that humanity must be more at the forefront of our environmental deliberations. Hannes Bergthaller et al. observe that those working in science-dominated fields "have come to recognize that scientific know-how by itself is insufficient to successfully address climate instability, soil erosion, freshwater shortages, and a host of other chronic social-ecological afflictions. Clearly, the ecological crisis is not only a crisis of the physical environment but also a crisis of the cultural and social environment—of the systems of representation and of the institutional structures through which contemporary society understands and responds to environmental change (or fails to do so: hence the crisis)" (262). Erle Ellis, a professor of geography and environmental systems, corroborates this observation. In a recent *New York Times* op-ed, he bluntly states, "Putting expert scientific narratives at the center of decision making, like 'nonnegotiable environmental limits,' rather than focusing on opportunities for collective betterment, has led only to increasing divisions over which experts to trust. If we are to continue improving the human world, while retaining the nature we love, it will be necessary to get beyond polemics and expertise, scientific or otherwise. In the end, it is people, and their institutions—not science—that will decide the future" ("Science Alone").

Despite our reliance on science and technology to assist us as we address complex environmental challenges, these areas cannot bear the sole responsibility, nor should they. And the fact is that while ample evidence exists for the extensive environmental challenges sweeping the globe today, the technical data communicating the reality of our precarious situation often fail to provoke meaningful action. All too often, such information is "delivered 'downstream' to a population of passive consumers" who are too overwhelmed by bad news, too indifferent as environmental collapse seems inevitable for them, or too ill-equipped to interpret the data in any meaningful way (Emmett and Nye 7).[1] The time has now come to make an important shift in our approach to solving environmental issues. Heeding Sörlin's call to establish alternative forms of "environmentally relevant knowledge," *Tributary Voices* examines a wide representation of texts for their values and beliefs that help us understand how we have arrived at our present situation and what potential solutions exist for establishing a more sustainable water ethic within the Colorado River Basin. Indeed, as Sörlin concludes, "It seems this time that our hopes are tied to the humanities" (788).

Citing both Bergthaller's and Sörlin's contributions to the emergence of the environmental humanities, ecocritic Ursula Heise clarifies the approach of this relatively nascent field, reaffirming the humanities' unique position in addressing environmental concerns and noting specific tools the field employs to bring together a variety of environmentally centered disciplines both in and out of the hard sciences.[2] In her introduction to *The Routledge Companion to the Environmental Humanities* (2017), she explains that the environmental humanities "envision ecological crises fundamentally as questions of socioeconomic inequality, cultural difference, and divergent histories, values, and ethical frameworks" and notes that because "scientific understanding and technological problem-solving, essential though they are, themselves are shaped by such frameworks and stand to gain by situating themselves in this historical and social cultural landscape" (2). Failure to perform such self-reflexive, disciplinary soul-searching to reveal the ways in which the humanities inform the sciences—an influence not often considered outside the humanities and typically addressed in the reverse form, as the sciences inform and shape the humanities—further alienates these fields and masks the ways in which they indeed inform one another. This divide is particularly evident in the following chapters where desert writing negotiates scientific and aesthetic knowledge in order to communicate the power and agency of desert water, the basin's tribes' understanding of water that embraces Western science and their own respective epistemologies, and John A. Widtsoe's mingling of science and his faith to address water stewardship. In sum, the environmental humanities—or more appropriately defined as the hydro humanities in regards to the current argument—are a necessary contribution to the broad and complex web of Colorado River representation and management. The hydro humanities' reliance on many different frameworks and theoretical approaches can help clarify the beliefs, assumptions, and values that ultimately shape perception and action about water use.[3] With this belief at the forefront of *Tributary Voices*, I now turn to a set of theoretical orientations and metaphors that give shape to the hydro humanities and establish a discourse ecology of water literacy, revealing the value of marginalized hydromythologies.

New Western Rhetoric, Bioregions, and the Watershed

Skeptical of the prevailing myths that construct a very unified and reductive view of western progress evident in works like Turner's frontier thesis and popular notions of the region, historians and literary scholars have, over a number of decades, recast the West as a much more complex and fraught space. The result of this ongoing renegotiation of place and culture has given birth to postwestern studies, New Western History, and, more recently, critical regionalism. The postwestern traces its roots to the 1970s, as Susan Kollin explains, wherein literary scholars began to examine works "that developed new interpretations of the West's past," which "provided viewers different viewpoints and loyalties from the usual fare" of heroic explorers, trappers, and cowboys (xii). A decade or so later, a dramatic shift within western American history further questioned traditional western values and ideals, leaving many frontier aficionados rankling at this "revisionist" history. Dubbed "New Western History" by individuals like Donald Worster and his fellow historians Patricia Limerick and William Cronon, among others, this approach positions the West in a new light, one typically far from the optimistic vision Turner's thesis had suggested. In Worster's words, New Western History "emerged out of the old clouds of myth and romance, and now seems for the first time honestly revealed" (*Under Western Skies* 5). While New Western History was quickly attacked by many in and out of the academy, it nonetheless began an important dialogue about our perceptions of the region and its human and nonhuman inhabitants.

Such critiques are particularly germane to the Colorado River Basin and the region's water history. As Limerick explains in *Legacy of Conquest* (1987), one of the foundational texts to New Western History, "People moved to arid and semiarid regions, secure in the faith that water would somehow be made available, then found the prospect of water scarcity both surprising and unfair" (43). Ironically, however, the reality of the situation did not prevent the garden myth from flourishing. Marc Reisner suggests, "Even as the myth of the welcoming, bountiful West was shattered, the myth of the independent yeoman farmer remained intact" (51). With this hope still in play, the arid West has seen more than a century of efforts to reclaim the desert and transform it into a well-watered oasis. Yet in light of today's increasing demands for a limited quantity of water, the perpetuation of this dream seems misguided, if not downright foolish.

A significant problem related to this inability to give up on the myth is the effect that propagating the myth has on those throughout the watershed. Since Turner's thesis envisioned a place of renewed opportunity and progress, the supposed savage cultures and landscapes of the region did not stand a chance in the face of Anglo progress that moved steadily westward, armed with enlightenment and civilization. Ultimately, Turner's thesis fails to account for the region's realities and complexities by promoting instead his belief in the nation's agrarian history.[4] Thus, for so many who have based their understanding of the West on Turner's observations, they have accepted a very reductive view of what the region is generally about and whom it is for. As a result, the land, its resources, and its Indigenous cultures have suffered mightily, as the nation has often looked to the West as the land of endless possibilities. "Quite simply," Worster explains, "the domination of nature in the water empire must lead to the domination of some people by others" (*Under Western Skies* 31). Such has often been the case with groups such as women, American Indians, Hispanos, Mexican Americans, and Mexicans throughout the Colorado's development over the past century or so. Yet while these tributary voices seem absent from the broader discourse about the river, a closer look reveals their ongoing presence that informs our understanding of the river through the voices' heterotopic nature to work within and against dominant narratives.

This need to acknowledge the value of lesser-known river perspectives draws strength from Limerick's second major addition to New Western History, *Something in the Soil* (2000), which reflects on the contribution this field has made to the cultural landscape of the region. Recognizing the public's poor reception of much of the early scholarship in this field because it supposedly "cast white men as wicked and demonic and envisioned the history of the American West as an unrelieved tragedy" (17), Limerick clarifies her position on western history through four key terms: "continuity," "convergence," "conquest," and "complexity." Limerick explains that the West is a place of continuity where the same practices at play before the closing in 1890 continue today, evident in such processes as ongoing boom-and-bust cycles in mining and other extractive industries (19). It is also a place of convergence, "one of the great meeting zones of the planet," where peoples of countless races, ethnicities, and religions interacted—a reality that challenges the belief that the West was just a place for white Americans to occupy (19). Conquest and complexity round out Limerick's description, suggesting that a simplistic, ahistorical notion of the frontier fails to acknowledge both the reality of widespread oppression of minority groups and the West's highly intricate past that is far from the naive escape to innocence suggested in Turner's work (20–21). Together these ideas aptly describe the Colorado River's past and present evident in the continuity of frontier optimism that influences water use and the region's development, the interaction of diverse cultures and beliefs about the river's value, and the oppression of peoples and resources within the watershed.

This emphasis on conquest and complexity is equally significant and closely related to the advent of critical regionalism and its reenvisioning of the American West. Literary scholars Krista Comer and Neil Campbell have provided important commentary on the New West that they approach through the postmodern shift. Comer contends in *Landscapes of the New West* (1999) that "landscape is not an empty field of vision... but rather a brimming-full *social* topography that creates and enacts the various cultural assumptions and power struggles of the age." Like Limerick, Comer is critical of past beliefs that landscape and the West in particular have been viewed as blank slates ready for the dominant culture's sole imprint. Rather, she recognizes the multifaceted and competing histories, philosophies, and practices that have shaped the land and suggests that "we might approach western landscape studies, and all of the social relations that landscape represents and negotiates, from a more nuanced and proactive vantage point" (13). *Tributary Voices* responds to Comer's invitation, as it encompasses a more comprehensive approach to the myriad discourses emerging from the watershed's diverse time periods and cultural groups.

Echoing a call to establish more productive perspectives by which to examine the West's landscapes as a site of critical discourse, Neil Campbell draws on Mikhail Bakhtin's notions of dialogism and polyphony to better understand how the American West operates as a complex social and physical space. Such perspectives lead him to view the "lived space of the West as a complex web of interconnections, competing discourses, and different 'voices' that together constitute the region" (10). He continues to explain that viewing the West through polyphony, this continual mingling of viewpoints, removes "boundaries constructed by narrow historiographical studies, dime novel stereotypes, and romantic visualizations of landscape and human action, by introducing new discourses in dialogue with the old" (11). Campbell's application of Bakhtin's ideas to the West provides a solid grounding for the importance of my notion of the Colorado's tributary voices that are continually in conversation with discourses of the past. Like rocks protruding from a stream that both shape the watercourse and are in turn shaped by the water, the tributary voices work within and against these past positions as they strive to leave their own imprint on the full body of river discourse.

Similar to the New Western historians and those promoting a critical regionalism of the West, I am equally interested in establishing new vantage points by which to examine the Colorado River Basin as a more complex discursive space. The history of the American Southwest is in many ways a tale of power struggles that have compromised the water rights and access to resources of Indigenous, Hispano, Mexican, and rural white communities, for example. At the same time, much of this dominance has come in the form of ideological and discursive power transmitted through narratives of individuality, progress, and science at the expense of those promoting community and faith. Thus, I posit that a New Western Rhetoric is necessary to address the very interdisciplinary and complex issues governing western water politics that often marginalize certain modes of communication and groups of people. I envision New Western Rhetoric as a special kind of discourse that acknowledges these disparate and conflicting regional perspectives from a wide range of backgrounds. However, while I am interested in the historical contributions a particular tributary voice makes to the river's timeline, I am more concerned with analyzing these voices' discursive strategies and appeals rooted in story, which forward their respective visions of the river and how such narratives engage, challenge, and even reject mainstream perspectives to posit alternative ways of living with the river. Therefore, while history plays an integral part in the delineation of this New Western Rhetoric, the primary contribution of this alternative approach is its emphasis on and critique of discursive activity rather than a celebration of a tributary voice's mere presence. From New Western Rhetoric's foregrounding of alternative narratives rooted within a quintessential western place, a discourse ecology of water literacy emerges that offers a more comprehensive and equitable representation of Colorado River use.

A discourse ecology of water literacy continues to take shape through ideas and metaphors integral to the development of place-based ecocritical studies that, in turn, are informed by ecology, geography, and other scientific fields. Among these is the bioregionalism movement and the related concept of the watershed. Bioregionalism emerged in the 1970s through the work of Peter Berg and Raymond Dasmann, who advocate for a reassessment of one's connection to a particular place that involves sociopolitical boundaries aligning with geographical features. This realignment recognizes that all entities, human and nonhuman alike, are part of the same community. They espouse a process of learning to "live-in-place" that "involves applying for membership in a biotic community and ceasing to be its exploiter" (36). For many ecocritics like Serenella Iovino, whose narrative reinhabitation is directly indebted to Berg and Dasmann's work, reinhabitation has become a goal to foster more sustainable connections to a place through place-based literature. Other ecocritics such as Tom Lynch observe, "By foregrounding natural factors—primarily the biotic community and watersheds—as the basis of a place's definition, bioregionalism posits that our sense of identity may be constituted by the characteristics of our residency in a larger community of natural beings—our local bioregion—rather than (or perhaps, more realistically, supplementary to) national, state, ethnic, or other more common bases of identity" (18). Unlike the rectilinear boundaries that delineate many western states (the Four Corners states are the perhaps the most obvious example), a bioregional orientation reenvisions the land and its resources as the primary factors by which to understand our connection to place and others.

Although cognizant of a region's diversity, a bioregional ethic embraces this difference as an enrichment of the geography of which it is a part. The end result of this practice, then, is what Gary Snyder, one of the great proponents of bioregional thought, observes: "There are 'classes' which have so far been overlooked—the animals, rivers, rocks, and grasses—now entering history" (41). Another way of saying this is that *where* we live has an immediate impact on *how* we live, since those who adopt a bioregional outlook are more attuned to their actions and their influence on the natural world and make necessary changes to ensure the prosperity of all within that biotic community. Besides the immediate value bioregionalism has in altering our relationship to place, it proves highly useful in rethinking the literature and other textual material that expresses our connection with a locale. As Lynch further suggests, "In addition to altering our sense of regionality, a bioregional approach is also especially valuable in the study of the literature of the American West for providing an alternative to the still pervasive trope of the frontier as a defining aspect of the region's culture" (26). Because bioregionalism demands a recognition of and a humility toward the natural world, it challenges traditional notions of place that position it as exploitable.

Bringing a bioregional approach to our understanding of the Colorado River and related western water issues is nothing new. In fact, one could argue that our present understanding of bioregionalism originated in response to the Colorado River well over a century ago. Having recorded the first passage down the Colorado from Green River, Wyoming, to the Grand Wash Cliffs that mark the end of the Grand Canyon in 1869, John Wesley Powell became an overnight hero as he brought the last "unknown" place in America to light. His subsequent journey down the river in 1871 and extensive travels throughout the American West from 1870 to 1879 as leader of a federal survey led Powell to promote a unique vision for the development of this arid region. Outlined in his *Report on the Lands of the Arid Region of the United States, with a More Detailed Account of the Lands of Utah*, which he submitted to Congress in 1878, Powell describes the land west of the 100th meridian that he believed could "eventually be rescued from their present worthless state" (*Arid Lands* 8). Noting the pervasive lack of precipitation throughout the region, Powell envisioned new land systems that would better adapt land-use practices to actual conditions rather than the current 160-acre plot granted to landowners under the 1862 Homestead Act. Drawing from his knowledge of New Mexican acequia culture and Mormon and Native irrigation practices (Lynch 45; Worster, *River Running West* 495; Stegner, "Thoughts" 50), Powell believed in a reorganization of the West into "hydrographic basins rather than by arbitrary political lines drawn on a map. Such basins... would be virtually self-governing and hence able to negotiate with other similar basins, as well as to control their own watersheds clear to the drainage divides" (Stegner, *Beyond the Hundredth Meridian* 315, 322). Powell looked to these self-sustaining, community-oriented groups as the arid West's best option for long-term development.

John Wesley Powell, *Arid Region of the United States Showing Drainage Districts.* "Eleventh Annual Report of the Director of the United States Geological Survey," pt. 2, "Irrigation: 1889–1890" (plate 69). Courtesy of Government Information Collection, Special Collections and Archives, Utah State University Libraries.

In his biography of Powell, Worster further elaborates on this revolutionary approach to establishing political entities based on topographical realities rather than more seemingly random survey lines. Of these geopolitical basins he writes, "The purpose now was to see the entire region as a mosaic of interconnected watersheds, as integrated units of water and land, not to deepen geological understanding so much as to guide settlement. Each of those carefully mapped and measured watersheds furnished the natural boundaries for a series of 'irrigation districts' into which settlers could come and work out their problems together" (*River Running West* 477). One can only imagine how the West would look today had Congress and the delegations throughout the region to which Powell promoted this idea accepted his watershed model. Yet those to whom he addressed this argument could not divorce themselves from the power of myths and the potential for profit. Thus, despite Powell's telling observation that "the lands which might thus be reclaimed are of greater extent than the amount which the streams can serve," politicians and boosters vehemently rebuffed his vision, placing their hopes in the garden myth rather than in one who had gained such an intimate and extensive knowledge of the arid regions through his explorations and associations with the region's Indigenous cultures (*Arid Lands* 17). While Powell's contemporaries rejected his progressive thinking, his ideas were not lost to history. As Worster observes, "In the years to come, practical men and women looking to create a new West...might reexamine the social and environmental ideals of John Wesley Powell, distilling out of them their democratic essence" (*Rivers of Empire* 332). It seems that such a day has come about, as numerous scholars across various disciplines have invoked Powell's model as a needed corrective to past thinking and practice.[5]

Indeed, a watershed model is particularly germane to how we think of and imagine the river along with our attempts to find workable solutions in light of the Colorado River's projected water shortages and escalating conflicts between cities, states, agriculture, industry, and environmentalists. As a particular kind of bioregion, and as Powell well understood, watersheds do not conform to political boundaries. In their most basic, topographical sense, a watershed is "the total land area that contributes water to a river...delineated by a ridge or drainage divided that marks the boundary of the drainage basin and can be easily identified on topographic maps. All surface water runoff below a ridge line will flow downhill within the watershed" (Cech 59). Watersheds can be relatively small, encompassing a particular drainage of a few square miles, or they may span large portions of continents, encompassing numerous states and other geopolitical spaces like the Mississippi watershed, the third largest in the world (65). And like bioregions, they are ripe for metaphorical application to New Western Rhetoric and the role that tributary voices play in reshaping our understanding of the discursive constructions of the river through literary and nonliterary texts.

In *Writing for an Endangered World* (2001), ecocritic Lawrence Buell claims that "'watershed' [has] become the most popular defining gestalt in contemporary bioregionalism, at least in the United States" (246). As such, numerous environmental writers have looked to watersheds specifically as a means of exploring new possibilities in ameliorating environmental problems. For Buell's own treatment of nature writing, he sees the value of what he dubs "watershed aesthetics" in their "respect for environmental health, equity, and self-restraint within particular bailiwicks, implying that cocoon-like fantasies of self-containment fail to grasp the complexity and extent of the imbrication of cultures and environments. For, in principle, 'watershed' ought to refer not only to a small, relatively finite unit but to a series of zones connecting local communities with larger stretches of continent" (263–64). As in bioregionalism, a sense of community cohesion and restraint is integral to the watershed model. Similarly, watersheds bring together culture and environment, providing a means of working around the nature-culture binary that has long vexed our thinking about human relationships with the natural world. However, more effectively than bioregionalism, at least in terms of thinking about the Colorado River, a watershed approach allows us to consider how water ties multiple bioregions and their related human components together. For example, the Colorado River watershed, as defined by the federal government, is composed of countless smaller watersheds consisting of its major and minor tributaries (for example, the Green, Gunnison, and San Juan Rivers) as well as numerous bioregions (the Colorado Plateau and Mojave and Sonoran Deserts) as it flows from its headwaters to the Gulf of California. But these watershed connections extend even beyond the watershed borders, as transbasin diversions move water from the watershed to places such as Albuquerque near the convergence of the Arizona–New Mexico Plateau and northern Chihuahuan Desert, Utah's Wasatch Front on the eastern edge of the Central Basin and Range, and Colorado's Front Range along the western fringe of the High Plains ("Level III"). Although unique in climate, flora, and fauna, each of these bioregions is nonetheless tied together by the Colorado River.

These realities demonstrate that the watershed is not just a hyperlocal heuristic by which to enact more sustainable living. While it can certainly function this way, the reality of current western water politics demonstrates that a watershed is an entity constituting local, regional, and even global interests. Thus, although Moab and Mexicali could not be more different from one another, they are united by the river that is the lifeblood for both communities. We can even extend this connection across continents, as farmers export alfalfa irrigated by the river to Japan, China, and Saudi Arabia to meet rising demands for beef in those regions (C. Turner; Markham). As such, the watershed model becomes a powerful "vision corrective" because it addresses very local concerns while at the same time extending its scope to include the realities of today's globalized world (Buell, *Writing* 246).[6]

A final articulation of the watershed model that moves us toward understanding how a discourse ecology of water literacy operates comes from Robert Adler, a lawyer, professor, and expert on Colorado River law and management. Building on the notion of connectivity inherent within watersheds, which so many other writers have acknowledged, Adler pushes the description further, suggesting the deep networks that watersheds make between different communities and the role the overall health of the system plays in restoring a particular place like the Colorado Basin. In *Restoring Colorado River Ecosystems* (2007), Adler writes,

> Watersheds are not just bodies of water, but are connected intimately with the entire associated land mass.... [R]iver systems are not just two-dimensional (linear and lateral) but four-dimensional in nature. They are longitudinal or linear (upstream-downstream); lateral (river floodplain to riparian zone and beyond); vertical (groundwater-surface water); and temporal (all three spatial dimensions change over time). Thus, for the Colorado as with all major rivers, the health of the river depends in part on what occurs on associate watershed lands. River restoration efforts cannot end at the water's edge, or even in the relatively limited riparian zone. We cannot view rivers as lines of water flowing downstream, disconnected from their sources of food and energy. In restoring the Colorado River and other water bodies around the country, we must consider connections in all dimensions. (77)

I quote at length Adler's consideration of a watershed as a multidimensional entity of various temporal and spatial relationships for its acutely relevant metaphorical potential in outlining the value of tributary voices, their narratives, and the broader discourse ecology of water literacy of which they are a part. Analogous to Adler's geographical definition of a watershed, a discursive watershed constructed by the many representations of the river embodies a similar multidimensional structure: They are often linear, as they follow western narrative structures; they are lateral, as some are disseminated further than others throughout the public sphere; they are vertical through their hierarchical nature, as some discourses take precedence over others as the "right" or "correct" representations of the river evident in the discussion about myths; and they are temporal in that they—like the river itself—change over time. Applying this four-dimensional watershed model to the literary and rhetorical representations of the river provides us with a clearer understanding of how literature and other textual productions construct the Colorado River. By recognizing and rhetorically listening to a greater breadth of discourses from within the watershed, we are better positioned to work toward improving the river's overall health for all dependent communities.

Mapping Metaphor Within the Realm of Rhetoric

Having outlined how New Western Rhetoric and the watershed model avail us of a more comprehensive assessment of Colorado River representation through their connections between disparate communities and epistemologies, I wish to conclude this chapter by examining how the watershed model functions more specifically as a metaphor to enact this discourse ecology of water literacy. Because the watershed is both physical and imaginative (built via language), another heuristic is needed to make clearer sense of how the Colorado River has been defined and represented over time. Since this project focuses on the imaginative creation of the watershed through various discourses, which each have their own set of common themes, discursive features, and rhetorical strategies in promoting a particular image and understanding of the river, I turn to Walter Beale's *A Pragmatic Theory of Rhetoric* (1987) as a system of rhetorical thought and action to organize the contributions of the tributary voices and to examine how they construct their vision of the river's significance to particular discourse communities through the (primarily) written texts they employ.

Beale's pragmatic theory revisits and amends rhetorical models by Aristotle and Kenneth Burke, as well as principles from speech-act theory that provide greater insight to written rather than oral discourse that has traditionally been the primary mode of inquiry among rhetoricians

(8–10). His theory rests on what he identifies as a "semiotic grammar of motives" or a "system" that "relates to human constructions of reality in general" (9). This system draws from Burke's pentad as articulated in his *A Grammar of Motives* (1945) to forward instead what Beale calls a quadrad, a four-pronged approach to understanding how discourse is motivated and how such discourse reflects a particular construction of reality. Beale points out that his semiotic grammar of motives "is not in itself a construction of reality but an objective model for talking about human constructions of reality" (10). Applied to the Colorado River and its watershed, this model helps initiate a discussion about how these tributary voices construct their own unique realities of the river and region and where their approaches and attitudes may align with or challenge the pervasive "reality" of what the river has meant in the past and what it possibly means today.

A fundamental part of Beale's discourse model is the "Motivational Axes," which consists of a vertical axis and horizontal axis, the two axes together plotting different philosophical positions about how language constructs reality and meaning. On the vertical axis the two poles correspond to how discourse functions to promote action and participation or contemplation and nonparticipation. On the horizontal axis discourse is referential or nonreferential (11). These two axes and the four poles that address how reality is constructed constitute Beale's quadrad, his "map to the entire range of linguistic and intellectual activity" (63). If we consider again the physical watershed that delineates all the water within a specific geographical space, Beale's philosophical and linguistic map contains the full range of written discourse about the river. In this sense, we can imagine the material watershed giving way to its metaphorical offspring—this discursive watershed through which flows a river of words. And while Beale's quadrad and grammar of motives are specifically focused on written expression, it works equally well to analyze the few visual genres (for example, short films) that appear later in this study. Thus, transposing Beale's discourse map onto Adler's multidimensional watershed creates a useful framework by which to analyze a range of discourse activity across variations in time and space and their influence within the public sphere. From here, if we overlay this new combined system of understanding discourse activity onto the geographical confines that demarcate the Colorado River Basin, we arrive at a comprehensive framework that connects the tributary voices' material, linguistic, and intellectual activity

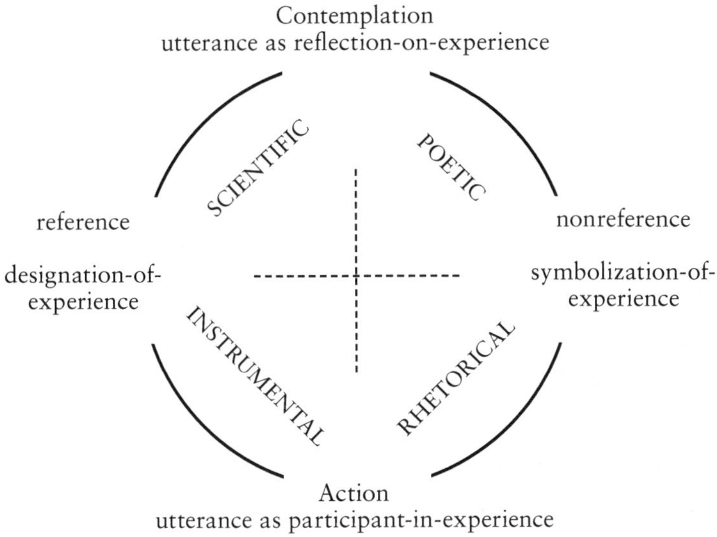

Walter Beale's *A Semiotic Grammar of Motives*. Redrawn from "A Semiotic Grammar of Motives," figure 1, from *A Pragmatic Theory of Rhetoric* by Walter H. Beale. Copyright 1987 by the Board of Trustees, Southern Illinois University. Courtesy of Southern Illinois University Press.

that defines the tributary voices' relationships with the river and other discourse communities.[7]

As we begin to plot the tributary voices' discursive activity onto this comprehensive watershed map, we should note the fluidity between poles that exists within written discourse. For Beale, "discourse [is] a complex human activity, to be understood in terms not of primary determinants but of ratios" (11). This is because, as Beale further observes, "no utterance...can call into play one of these axes without simultaneously invoking the other" (64). Consequently, when a text invokes a more contemplative orientation, such contemplation either is based on a referent noted by an observer—much as a hydrologist does when analyzing stream flows to then formulate various conclusions about their impact on native fishes, for example—or alternatively might be more closely oriented to a less objective and more nonreferential entity, to use Beale's words, as when a poet sings the praises of a beautiful mountain view. Together, these combinations or "ratios" of how meaning is constructed give rise to the classical aims of discourse: scientific, poetic, and rhetorical. To this list Beale adds instrumental to better account for the proliferation of written discourse that exists today and that was of lesser significance in Aristotle's world (90). Thus, the four poles of the motivational axes (contemplation/action, reference/nonreference) give rise to these four aims of discourse (scientific, poetic, instrumental, and rhetorical), which embody portions or combinations of these philosophical orientations. As such, scientifically motivated discourse is a ratio of contemplation and reference, poetic discourse combines contemplation and nonreference, rhetorical discourse negotiates nonreference and action, and, finally, instrumental discourse blends both action and reference (11, 64).

Considering the Colorado River watershed's past and present representations in light of pervasive myths, these aims help demonstrate the range of voices and their associated intentions for river use, their interconnectivity with one another, and, in some cases, the manner in which they challenge more mainstream discourses. To this end Beale reminds us, "Any single work of discourse may contain or exhibit a range of motives, of course—some of them explicit in the discourse, others implicit in the situation, others transparent only to a subset of the audience, and others immediately accessible only to the author" (95). Because it is neither possible to re-create the exact situations that motivate my tributary voices' discursive acts (although the intentions in AEURHYC's document and the Ten Tribes Water Study are more readily apparent than Childs's nature writing, for example) nor possible to fully understand how an author and audience intended and perceived such motives, my analyses focus on the motives that emerge from the texts themselves.

To negotiate and examine these various motives, their respective aims, and their relevance to how my tributary voices discursively construct the river and watershed, I consider their primary objectives and the genres that most typically represent them. In the case of the scientific aim, its main objective is "the discovery, construction, and organization of knowledge, particularly in those areas or subareas in which facts, classifications, and general laws can be verified by rational and empirical procedures, as opposed to the values and loyalties of communities." Beale cites such specific genres as reports and theoretical and philosophical treatises as emblematic of the scientific aim. The poetic focuses on "the construction of an object of enjoyment and reflection, using the materials and resources of language" found in such genres as "poems, stories, and novels." Along with the scientific and poetic aims of discourse, the rhetorical and instrumental flesh out the rest of discursive activity humans use to convey versions of reality. The rhetorical seeks "to influence the understanding and conduct of human affairs" and "operates typically in matters of action that involve the well-being and destiny of communities" and "in matters of value and understanding which involve the communal or competing values of communities." Beale lists such texts as "public resolutions and declarations of competing groups within a community" as examples. Finally, Beale defines the instrumental aim through its "governance, guidance, control, or execution of human activities. It includes such specific products as contracts, constitutions, laws, technical reports, and manuals of operation." Again, it is worth reemphasizing that these descriptions are "not ideals; they are norms of activity" (94).

Because these distinctions between modes are not absolute, they may perform in unexpected ways. For example, Childs's nature writing, the Ten Tribes Water Study, and John A. Widtsoe's reflections on western reclamation embody what may be described as both scientifically and poetically motivated discourse, whereas Patricia McCaren's Grand Canyon boating narrative and Leslie Marmon Silko's *Gardens in the Dunes* appropriate the genre forms of the western and Victorian novel, respectively, to critique and challenge their themes and, ultimately, the expectations of these forms of poetically motivated discourse. According to Beale, "An individual work of discourse succeeds partly by conforming to the norms of a given rhetorical genre, partly by transcending them, sometimes even by flouting them" (115). The success then of these tributary voices' texts to our present concerns regarding the river depends upon their ability to appropriate conventional genres and the typical discourses they embody and then to forward their own version of what the river and watershed mean.

Whereas the scientific, instrumental, and rhetorical aims have long dominated Colorado River discourse through the countless reports, laws, policies, procedures, regulations, and speeches that govern water use, my interest is primarily within the poetic and how narrative—as a foundational aspect of poetic discourse—transcends this aim to shape texts that initially appear to reflect more scientific, rhetorical, and instrumental purposes. As rhetoricians James Phelan and Peter J. Rabinowitz explain narrative, it is "itself an event . . . *a multidimensional purposive communication from a teller to an audience*." As such, the authors maintain that "the elements of any narrative (e.g., character, setting, plot structure) are shaped in the service of larger ends" (3). Phelan, on an earlier occasion, clarifies that "narrative is not just story but also action, the telling of a story by someone to someone on some occasion for some purpose" (*Narrative* 8).

As we will see with these tributary voices, narrative does not exist purely for narrative's sake. Rather, narrative operates not only to entertain, as story often does, but also to communicate important lessons, underscore specific values, challenge long-held beliefs, and invite action in order to cause us and their various audiences to pause, reflect, and weigh the significance of their perspectives. Therefore, an important component of the present study is to identify and examine the value of "departures" or cross-currents that tributary voices embody in light of their "convergences" or confluences with the pervasive mainstream discourses that have dictated how we have viewed the river over time. Their use of narrative, as it engages other discourse aims, allows us to see how tributary voices utilize the dominant aims while incorporating the poetic to facilitate contemplation that can ultimately lead to action along scientific, instrumental, and rhetorical lines. This, in turn, enables us to create new stories about the river and watershed that will better equip us to face tomorrow's challenges. By examining representative texts from a number of tributary voices that align to varying degrees with Beale's four discourse aims, the Colorado River watershed becomes a complex site where multiple discourses, genres, and attitudes intermingle. As they interact with prevailing beliefs about how to imagine and use the river, a broader understanding emerges to a larger set of constituencies of what the Colorado River means physically, discursively, and ideologically.

Having established the key theories, frameworks, and models by which to interrogate and illuminate the tributary voices, I wish to address, in conclusion, a fundamental question that emerges out of the previous discussion: Having identified these marginalized watershed perspectives and the various stories they tell, how do they make a difference in our attitudes and practices regarding the river? That is, how do story and contemplation eventually move us toward actions where we will make changes in behavior? While there are a variety of responses to these questions, I offer one strategy that emerges from the way we listen to the narratives and the types of stories we choose to tell with our newly acquired knowledge. Krista Ratcliffe advocates "rhetorical listening," which she defines as "a stance of openness that a person may choose to assume in relation to a person, text, or culture" (17). The purpose of this approach, as Ratcliffe further explains, "is to cultivate conscious identifications in ways that promote productive communication, especially but not solely cross-culturally" (25). While initially addressing gender and racial inequities, Ratcliffe's framework proves useful as we seek to understand and negotiate stakeholder positions that often run in conflict with one another. Thus, this pursuit of identifications—the commonalities and shared interests and even the differences stakeholders have with one another—serves as an important strategy by which to engage tributary voices.[8] Rhetorical listening essentially opens up a space in which we contemplate other stories and relationships to the Colorado with the goal of enhancing our understanding of where we and others stand on particular issues, especially when common ground is hard to come by. In fact, recognizing and allowing for these differences is essential to rhetorical listening, as such awareness fosters a more complete picture of the ways in which a stakeholder views the river. In the case of women's boating narratives, and the voices from Latina/o, Indigenous, and religious communities, allowing for different cultural logics from the mainstream is essential to facilitating a discourse ecology of water literacy. A cultural logic, as Ratcliffe explains, is "a belief system or shared way of reasoning." By identifying the various cultural logics at play within this study, "listeners may still disagree with [the tributary voices'] claims, but they may better appreciate that the other person [or group] is not simply wrong but rather functioning from within a different logic" (33). Seeking to understand these alternative perspectives, we are in a better position to begin to work toward solutions that are beneficial to all parties. In the end, rhetorical listening creates "an accountability logic" where "all people necessarily have a stake in each other's quality of life" (31).

Perhaps a simpler yet equally effective way of understanding how this accountability functions within the Colorado River Basin is to consider Phillip Fradkin's reorientation of our own positionality within the watershed and our claims to what we believe is "ours." In his extensive travels throughout the region, Fradkin observes, "But listening to and seeing how people used the river the Colorado became a series of segments, each labeled 'mine.'... One neighbor's vision rarely penetrated into another's backyard with any great degree of compassion.... To listen to the voices, one would never suspect they were connected to the same river, or that the Colorado led to Mexico and the expectations of its multitudes, or, after being last used in that country, that it no longer emptied into the gulf" (316–17). Most significant in this statement is the role that a critical listening to voices plays in gaining a "watershed" vision of the river. Fradkin's view of the river is radically different from those perspectives that are limited by their own small stretch of the river or its countless tributaries. Indeed, while most people in the Upper Basin think little of their downstream neighbors and vice versa (except, perhaps, to blame them for taking too much water), the decisions made within each basin regarding the management of the river can have significant impacts elsewhere. Yet because Fradkin had traveled from the headwaters to the mouth and taken time to listen to the stories of those all along the river, he is able to patch these segments together to form a more comprehensive view, one less inhibited by the biases and ignorance of those who lack such vision. Fradkin's awareness of the entire watershed, evident as he brings together disparate perspectives, aligns with the previously discussed theoretical positions that call for a more complex, multipronged approach to understanding the American West. While few will have the opportunity to travel the watershed as extensively as Fradkin, we can avail ourselves nonetheless of the various perspectives that constitute the wide discursive watershed. We can take time to listen, process, and understand narratives that might run counter to those told within our own areas of the watershed or within the specific discourse communities to which we belong. We might be more open to what the agricultural community has to say or the growing cities, to the stories the Diné recount, or to those advocating in-stream flows for endangered species.

As we now turn to a selection of tributary voices that reflect differences in time, geography, gender, race, and religion, I invite you to rhetorically listen, to consider how a discursive ecology of water literacy unfolds through New Western Rhetoric, the watershed approach, and Beale's motivational aims. Together they broaden the range of perspectives invested in the river's past, present, and future and how their use of story—reflected in a host of genres—encourages our narrative reinhabitation of the Colorado River Basin. As we bring this extensive and rich coalition of theoretical viewpoints to bear upon the following texts, we recover valuable perspectives that should have been part of the region's representation all along. From these we can draw needed inspiration by which to inform Colorado River governance moving forward with the hope that while turbulent times are ahead, we can navigate these challenges with a greater set of options and strategies to ensure greater water equity for all.

CHAPTER TWO

Science, Poetry, and Paradox

Bridging Divergent Epistemologies in Colorado River Desert Writing

In still another respect the Colorado basin offers a peculiar paradox. It contains one of the greatest river systems in America; and at the same time this whole vast drainage basin coincides with the Great American Desert.
—Frank Waters, *The Colorado*

In the opening pages of *The Secret Knowledge of Water* (2000), Craig Childs, a modern-day desert explorer, describes a journey to a southern Utah slot canyon in full flood stage. As the deluge subsides, Childs enters the gorge to examine the water, and it is within the depths of this chasm that he reveals water's unique and perplexing presence in the desert. With the "sour taste of dead animals and fresh mud" on his lips, he explains, "there are two easy ways to die in the desert: thirst or drowning. This place is stained with such ironies, a tension set between the need to find water and the need to get away from it. The floods that come with the least warning arrive at the hottest time of the year, when the last thing on a person's mind is too much water" (xiv). Musing on the consequences of too little or too much water, Childs invokes a central theme of the literature about the deserts of the American Southwest: paradox.

No stranger to this motif is Edward Abbey and his text *Desert Solitaire* (1968). Explaining his rationale for taking a job at Arches National Monument, he states, "I am here not only to evade for a while the clamor and filth and confusion of the cultural apparatus but also to confront... the bare bones of existence, the elemental and fundamental, the bedrock which sustains us. . . . I dream of a hard and brutal mysticism in which the naked self merges with the non-human world and yet somehow survives still intact, individual, separate. Bedrock and paradox" (7). Retreating from the constricting confines of society, Abbey finds his paradise in a remote corner of a southwestern desert where his imagination and zeal for freedom flourish. Grappling with notions about life and death, the real and the abstract, Abbey establishes his work within what David Pozza describes as "a philosophy of paradox" (14), one through which the author "gains a broader perspective on existence where inconsistencies and incongruities naturally abound" (9). Abbey's evocation of paradox places him in a long tradition of desert writers who marvel at the desert regions where life and death stand in stark contrast, where the land itself seems to contradict and challenge Anglo-American expectations of how the land should appear.

Of course, paradox as a literary device is not solely characteristic of these two authors and their respective works of desert literature. Cleanth Brooks argues in "The Language of Paradox" more than a half century ago that this motif plays a key role in the poet's efforts to unify what may initially appear as contradictory emotions, ideas, and symbols (17). Of the contempt that many have given to paradox in relation to poetry, Brooks notes that critics have regarded this motif as "the language of sophistry, hard, bright, witty" rather than the "language of the soul." He further suggests, "Our prejudices force us to regard paradox as intellectual rather than emotional, clever rather than profound, rational rather than divinely irrational," and that under such preconceived notions, "it is the scientist whose truth requires a language purged of every trace of paradox" (3). However, as I contend throughout this chapter, it is precisely the scientist who, like the poet, eventually comes to rely on paradox to represent both the desert's otherness from human existence and the connections between humans and the desert that shape and influence both entities.

For Childs, Abbey, and countless other writers, the desert is a place of contradictions that elicits a host of responses as desert travelers try to understand the landscape, its processes, and its inhabitants. In their respective works on the American desert-writing tradition, scholars Peter Wild and Patricia Limerick trace these apparent contradictions expressed within the desert narratives by Spanish explorers and missionaries, Anglo scientists and adventurers, and more contemporary nature writers.[1] Wild suggests, "Over the last hundred years or so, we have embraced a huge set of contradictions. Our culture has turned the desert, as if it were a limitless, exotic putty, into just about anything people want it to be" (*Opal Desert* 3). To make sense out of these literary unconformities, which reflect the Colorado's own myriad representations, Limerick categorizes the outcomes of such encounters in three distinct but related views toward deserts that derive their power from pervasive myths about the fabled Great American Desert (*Desert Passages* 3). These include "attitudes toward nature as a biological reality in human life—vulnerability to hunger, thirst, injury, disease, and death . . . as an economic resource—a container of treasures awaiting extraction or development; and as an aesthetic spectacle." Moreover, although Limerick acknowledges that "the phrase *attitudes toward nature* throws a thin cloak of simplicity over a mass of complexity," this unifying categorization of key themes and developments proves extremely useful in approaching a large body of work that represents significant leaps in time and space (6).

As ongoing drought and increased development and urbanization further stress the Colorado River Basin's limited water resources, we would be wise to take a closer look at this "thin cloak of simplicity," for, as Limerick argues of such representations and those who create them, "there is room to acknowledge subtlety, contradiction, and paradox" (9). By casting a more critical eye upon the texts representing the American deserts, we discover how the attitudes about the desert might, through the irony, paradox, and wonder they express, suggest alternative ways of knowing and acting toward these ecosystems and particularly those within the Colorado River Basin. In the case of Childs's desert travels, paradox defines not just what he observes but also the type of discourses he employs as he seeks knowledge. Therefore, Childs's use of paradox creates an alternative vision of these bioregions within the broader Colorado River Basin shaped less by an attitude projected toward nature than one *open* to nature. This reorientation endows the desert with agency or "speaks back" to the writer. Ultimately, this vision enacts the "voice of nature," which, as Matthew Teorey explains, is the "nonpersonified authority as a speaking subject that communicates, in its own way, to all living beings, including humans" a more complex view of the desert and its water (2). *The Secret Knowledge of Water* is therefore Childs's account of how the desert has instructed him and how, when confronted with paradox on an observational and experiential level, he has to rely on different discourses to make sense of desert phenomena. The treatment of the desert's indifference to humankind and its pattern for coping with aridity ultimately mold Childs's and other desert writers' thinking to offer up new ways to engage the watershed's deserts.

To examine how paradox defines this rich literary tradition, I draw upon Walter Beale's grammar of motives and ecocriticism's materialist turn to consider how this tradition engages both scientific and poetic discourses to make sense of the desert. Beale defines scientific discourse as "the kind…whose primary aim is the discovery, construction, and organization of knowledge, particularly in those areas or sub-areas in which facts, classifications, and general laws can be verified by rational and empirical procedures." To chart this relationship, I expand my analysis of desert discourse beyond the typical genres such as "reports of historical, statistical, field or laboratory investigations" that typically define scientific inquiry evident in the region's historical treatments. Poetic discourse, on the other hand, focuses on "the construction of an object of enjoyment and reflection, using the materials and resources of language" found in such genres as "poems, stories, and novels" (94). I add to this rhetorical framework Jane Bennett's work on vibrant matter, which asks us how we might better address our environmental crises if "we gave the force of things more due" (viii). Together, this rhetorical mapping and consideration of matter's agency allow us to read this tradition and particularly Childs's construction of the Colorado Basin's deserts as physical and philosophical contact zones where we confront the borders that mark epistemological systems, the divisions between the human and the nonhuman, and the discursive and disciplinary strategies and positions used to evoke the mystery and wonder of these bioregions. To illuminate these relationships, I turn briefly to the explorers and scientists who provide the first written expressions of the desert and then consider the influence their observations have on some of the most prominent desert naturalists and nature writers of the twentieth century such as Mary Austin, Ann Zwinger, Joseph Wood Krutch, and Edward Abbey whose works engage in an "eco-dialogue" with their surroundings (Teorey 6). I then read Childs's *The Secret Knowledge of Water* through these lenses and this genealogy to rethink our relationships with aridity and the vast spectrum of matter (human and nonhuman alike) shaped by this reality with the broader goal of identifying how we might more effectively imagine and inhabit the watershed.[2]

The Watershed's Explorer-Scientist Tradition

In the mid-sixteenth century, a number of Spanish explorers entered the watershed's deserts to provide the first written accounts of this region. Alvar Núñez Cabeza de Vaca and his three traveling companions are credited with being the first non-Indigenous people in the watershed following their storied journey from the coasts of western Florida to present-day Texas and the American Southwest before heading south into Mexico City, where they arrived in 1536 (Lavender, *Colorado* 14). Drawn by the rumors of massive wealth to be found in the region, subsequent expeditions penetrated farther into the deserts of what was then Spain's most northern territories in the New World. Under the leadership of Cortés and Coronado, the first accounts of the Colorado River came from Francisco de Ulloa, Hernando de Alarcón, and García López de Cárdenas who probed the coastline of the Gulf of California at the river's mouth and made their way overland to the Grand Canyon. The geographic and cartographic details gleaned during these initial forays into the region were of such little importance or were so imprecise that Juan de Oñate's report of his journey in the first years of the seventeenth century from the San Juan River Basin to the Gulf of California influenced mapmakers at the time to label the peninsula, previously noted as such by Alárcon, as an island once again (21). Not until the eighteenth century with the expeditions of Jesuit and Franciscan missionaries did the region's geography become more standardized. Eusebio Francisco Kino as well as Francisco Atanasio Domínguez and Silvestre Vélez de Escalante loom large in the history of the watershed as their travels throughout the Sonoran Desert—in Kino's case—and Domínguez and Escalante's travels throughout the Colorado Plateau country brought greater knowledge to Spain of its lands.[3]

In the nineteenth century, the first Anglo-American settlers made their entrance into the watershed's deserts, thereby extending the scientific knowledge of the region beyond general observations of topography and biology. Capitalizing on Jefferson's purchase of the Louisiana Territory that extended the nation's western border to the edge of the continent's deserts and what was at one time Spain's and then Mexico's sparsely populated northern borders, numerous mountain men and adventurers entered the deserts searching out their own dreams. Historian William Goetzmann dubs this period of American encroachment and entrance into this territory as the "Rediscovery of the Southwest," since those who made their way into this region did so without the accumulation of three centuries of knowledge by the Spanish explorers who had previously mapped much of the region (38). Without access to the historical records that previous explorers had produced—many of them housed in archives in Europe and Mexico—these newcomers were not privy to this knowledge. Thus, for the mountain men, traders, and ensuing soldiers, explorers, and scientists who would follow, "the essential task confronting them was one of rediscovery, of regaining the knowledge that had been lost" (39). This recurring pattern of discovery and loss and the subsequent desire to reconnect the lost knowledge become a key topos or commonplace in the desert literature of writers like Childs that allows for science and scientific discourse to take the primary role in shedding light on what has proved to be an enigmatic landscape.[4]

While knowledge about the watershed's deserts waxed and waned over centuries, this changed with Zebulon Pike's 1806 expedition to explore the southern portions of the newly acquired Louisiana Territory and the subsequent government-sponsored expeditions in the following decades. Although his travels never carried him into the Colorado Basin, his observations of the plains and the Arkansas and Rio Grande river drainages influenced how later generations viewed the Colorado watershed. In his February 5, 1807, entry, he likens "these vast plains of the western hemisphere...with the sandy deserts of Africa, for I saw in my route in various places, tracts of many leagues where the wind had thrown up the sand, in all the fanciful forms of the ocean's rolling waves, and on which not a speck of vegetation existed." Viewing the region as ultimately worthless in terms of its overall agricultural potential, as had the Spanish padres three decades before, he nonetheless finds the stark aridity something to praise, as America now had its own exotic deserts to rival those of more famous climes.[5] Finding the region utterly alien to his eastern sensibilities, Pike invokes "fanciful forms" of sand like the waves of the sea, which thereby infuse his work with a poetic strain that only adds to the mystique of the arid West (230). Confronted with such unfamiliar sites, Pike underscores the complexity of the discursive constructions of the region as they move between actuality and hyperbole, verity and myth.

Beginning in the 1840s, the federal government sponsored a number of expeditions to the nation's ever-expanding territory. A decade later, the Topographical Bureau charged the Corps of Topographical Engineers, the outfit primarily responsible for opening up the American West to development and settlement, with cataloging "the plants, animals, Indians, and geological formations of the country traversed" (Goetzmann 303). Seeking to advance knowledge in a variety of fields, including geology, botany, zoology, and ethnology, the bureau hoped to create "the broad outline, the comprehensive catalogue that was basic to any long-range scientific consideration of the region" (329). Such a far-reaching approach sets the stage for the desert writing in the following centuries, as this expansive understanding of science merges with more poetic responses to the desert, thereby opening the door to paradox as a rhetorical device and a broader range of discourse used to express one's engagement with the region. As historians have observed of science, exploration, and literature, these "were areas of activity that, while largely distinct from each other, were not always wholly separate or unitary fields" (Fulford, Lee, and Kitson 2). Perhaps the best example of this hodgepodge of disciplines comes from the famed German scientist Alexander von Humboldt and his work in the Southern Hemisphere. It is here, as Donald Worster explains, that Humboldt brought together a host of fields like botany, geography, and the emerging science of climatology to understand the interconnections between plants and their respective habitats. At the same time, however, this desire to move toward a sense of "truth" about the natural world was equally infused with an aesthetic of nature typical of the Romantic worldview (*Nature's Economy* 135–36). This all-encompassing view of science and nature enriched by various philosophical traditions and discourses proved highly attractive to his contemporaries. However, not yet willing to divorce objective scientific inquiry from their desire to see a broader harmony and order in nature, these writers' "work was still philosophical and metaphysical rather than scientific in its emphasis" as they searched for "unity, totality, and oneness with the whole 'Kosmos'" (Goetzmann 329).

Lieutenant Joseph C. Ives's expedition up the Colorado River with the Topographical Engineers in 1857–58 echoes the era's various responses to nature and science. Accompanied by a host of scientists, Ives attempted to navigate the river by his steamboat, *The Explorer*, in hopes of establishing a workable trade route to the settlements in New Mexico and Utah.[6] Advancing up the Colorado from its mouth as Alarcón had done three centuries earlier, Ives reached only as far as Black Canyon on the western end of the Grand Canyon, where he left his ship to proceed on land. He concludes that the region, from "a scientific point of view," is "of the highest interest." Yet he also praises the "strange sublimity" that "is perhaps unparalleled in any part of the world" (5). Thus, while the report is replete with information on hydrography, geology, botany, and zoology penned by Ives and a variety of experts, it also includes sketches and descriptions of the unique geology influenced by European Romanticism. When presented with a spectacular view of the "Big Cañon," Ives notes how his topographer, F. W. Egloffstein, turned to his sketch pad to create a depiction that "does better justice than any descriptions can do to the marvelous scene." Steeped in the Romantic notions of the sublime, which dominated nineteenth-century American thought about nature, Ives and his fellow scientists represent their surroundings in a unique blend of detached scientific inquiry with representations of nature that emphasized its divinity, its wonder, and often its terror. Despite the scientific and aesthetic wonders of the Grand Canyon, Ives finds the region ultimately worthless. He concludes, "Ours has been the first, and will doubtless be the last, party of whites to visit this profitless locality. It seems intended by nature that the Colorado river [*sic*], along the greater portion of its lonely and majestic way, shall be forever unvisited and undisturbed" (110). Little did he know that the "strange sublimity" of this "lonely and majestic" place would continue to attract explorers and scientists and eventually millions of visitors who rely on scientific and poetic discourse to make sense of this most iconic of desertscapes.

With the expeditions of the 1860s, the Romantic vision of the earth and its processes gave way to a more specialized, complex understanding of the region (Goetzmann 356). Yet, as is clear in the following example, old ideas do not easily disappear. Chief among these new explorer-scientists, and certainly the most famous and celebrated within the watershed, is John Wesley Powell. A self-appointed naturalist, Powell follows in the rich tradition of European and American naturalist-scientists who turned away from Christian theology to explain the earth's origins (Worster, *River Running West* 60). According to Worster, these men envisioned a world that "followed its own laws without any need for immediate or direct supernatural interference," believing that "science alone could explain those laws" (63). Having completed two journeys down the Colorado in 1869 and 1871–72 along with more extensive surveys throughout the Intermountain West and especially Utah, Powell tested his hypotheses about how the land was formed and how people could reasonably develop a region that seemed to all but resist widespread manipulation.

The findings of these activities resulted in a number of publications, including his famous river account captured in *The Exploration of the Colorado River and Its Canyons* (1895) and *Report on the Lands of the Arid Region* (1878). These texts reflect Powell's reliance on a number of scientific advances in a variety of fields to process his surroundings, as they are replete with his hypotheses and conclusions about the region's geology, hydrology, biology, and human presence. However, Powell's work—particularly his river narrative—also reflects a clear Romantic strain, as he speaks of the canyons' sublime precipices as Ives and others had done. As Worster explains of Powell, "Sometimes he seemed to be almost completely rooted in the earth—rooted in facts of soil and water, in the problems of getting a living from the land. At other times, however, his head was way up in the clouds, looking for the unseen—the laws that governed the evolution of life" (*River Running West* 549). Caught between empirical observation and philosophical musings deriving from Darwin's nascent theories about nature's origins and the earlier, more established Humboldtean science, Powell's observations and those of his fellow explorer-scientists give voice to the tensions within the scientifically motivated discourse of the period. As their observations and representations of nature shift between a more objective, even pessimistic, realism and Romantic idealism, these men ensured that others would view the natural world through a similarly complex lens. In the case of the watershed's deserts, three centuries of reports have depicted these places through a wide range of reactions, demonstrating that they cannot be pinned down into a nicely packaged and easily quantifiable whole. Vilified on one hand for their inability to provide food, water, and wealth while celebrated on the other for their sublime landscapes and potential for colonization and development, the deserts have proved highly contradictory for those searching to make sense of these complex landscapes.

From Natural History to Nature Writing

With the deserts mapped and its species cataloged, the days of the traditional explorer-scientist faded into history, or so it seems. While the Corps of Topographical Engineers no longer leads the charge in unveiling the region's secrets, a new type of desert-explorer has emerged to continue to shed light on humanity's connection with the nation's most arid regions: the nature writer. Drawing on the natural history tradition that influenced the United States' earliest explorers and scientists, these writers entered the watershed's deserts in search of recovering the "dry knowledge" of centuries past and expanding our understanding of these regions.

Emerging during the Romantic period through the work of Gilbert White, the English parson whose *Natural History of Selbourne* (1789) is the foundational text in this field, natural history writing has had a profound impact on how we view both nature and science. In his study of the development of ecological ideas in western thought, Worster looks to White and the particular cultural conditions shaping his study of Selbourne's flora and fauna. Influenced by the Industrial Revolution and increased specialization in the sciences (*Nature's Economy* 18), White and those who would follow embraced what Worster defines as an arcadian model, one in which "ecological study is the means by which the naturalist establishes communion with nature: it becomes an innocent pursuit of knowledge about one's neighbors, an integral part of the curate's devotional life, excluding in no way a sense of piety, beauty, or humility" (11).

As an offspring of the Romantic period, natural history sought to reconnect man and nature through a unique fusion of science and emotion, an approach embraced by Humboldt and popularized by countless other naturalists. This particular vision of man's relationship to nature would only strengthen after Darwin's publication of *On the Origin of Species* (1859), which challenged the orderly, holistic science that the naturalists working in the tradition of Gilbert White espoused. "One of the central burdens of the natural history essay," Worster explains, was to "find an alternative to this cold science [of Darwin]—not by retreat into unexamined dogmatism, but by restoring to scientific inquiry some of the warmth, breadth, and piety which had been infused into it by [White]" (*Nature's Economy* 17). The purpose of this particular approach is, as Worster further claims, "twofold: first, to endow each creature with a freedom of will and action that would defy analysis by chemistry and physics; and second, to study all nature as a single integrated unity, held together by a rather mysterious organizing force" (18). By today's standards, such an approach to understanding nature seems both archaic and naive, especially as the naturalists working in this tradition viewed the natural world through a highly moralistic lens.

Yet a closer look at this genre's development over time reveals that such ideas are not easy to discard. According to Thomas Lyon, the nature essay is "one of the most stable genres in literature" (Introduction 1).

Uniquely motivated by scientific and poetic discourses that blend inquiry and ethics, we see an emerging tradition beginning with Romantics like Wordsworth and continuing through contributions by transcendentalists Emerson and Thoreau, early wilderness advocates John Burroughs and John Muir in the late nineteenth century, and more recent writers like Aldo Leopold, Rachel Carson, Annie Dillard, and Barry Lopez. While all of these authors' works embody their own unique ways of approaching nature and humans' relationship to it, they nonetheless regularize a genre that relies on first-person interaction and observation to draw larger conclusions about how society thinks and acts toward the natural world.

In *This Incomparable Land* (2001), Lyon provides a particularly useful schematic for understanding the many valences of this genre. Creating a "taxonomy of nature writing"—a method quite apropos to this discussion as he draws on science's penchant for classification—Lyon outlines a range of nature writing that extends from technical field guides and natural history essays on one end of the spectrum to the other, where nature's emphasis gives way to the author's own experience (22). Regardless of where one chooses to classify a text within this taxonomy, there is a shared sense of the broader genre's overall purpose. For Lyon, "the fundamental goal of the genre is to turn our attention outward to the activity of nature." The result of this action is that there arises "a clarifying of perception inherent in this refocusing, which opens up something like a new world," a world that "may eventually be seen as a more important discovery than the finding of new lands" (25). This approach defines contemporary desert authors as they build upon the explorer-scientist tradition to offer their own perspectives on desert travel. While they do not discover new territory per se, they nonetheless introduce the reader to relatively uncharted imaginative terrain through their unique approach to "harmonize fact knowledge with emotional knowledge" (Lyon, Introduction 3). This mingling of disparate forms of knowing, coupled with the casting of old knowledge—that which was established by earlier explorers and inhabitants of the region—within the new knowledge of his own observations and scientific pursuits, brings paradox once again to the forefront of desert writing as a powerful means of reconciling the region's complex physical and imaginative realities.

The purpose of this unique epistemological approach of bridging fact with emotion through scientific and poetic discourse is to create what Lyon has called "an entire psychic reorientation" (4). By incorporating scientific "fact" alongside an author's aesthetic responses to his surroundings, Lyon argues that nature writing seeks to challenge the human-nature duality that has persisted since the Enlightenment and that gained much momentum during the period in which White first turned his attention to the nature of his village. Ironically—or, perhaps more appropriately here, paradoxically—this shift toward a more ecocentric view often comes through an author's foregrounding of personal experience and the emphasis of the self. Ecocritic Scott Slovic, one of the first scholars to examine the implications of this trend in nature writing, comments on its proliferation among numerous nature writers like Annie Dillard, Edward Abbey, Wendell Berry, and Barry Lopez. Of them he suggests that they "are not merely, or even primarily, analysts of nature or appreciators of nature—rather, they are students of the human mind, literary psychologists" (*Seeking Awareness* 3). Thus, through their observations of the natural world, the nature writers embark on a journey to remap the interior landscape, to posit alternatives to anthropocentrism. If the intent of nature writing is to encourage what Lyon describes as reorientation, then it seems, perhaps surprisingly, that a strong human presence in the form of the narrator is necessary. Through the narrator's experience, the audience vicariously experiences the natural world created within the text and thinks through the ethical questions posed as the author engages with nature. As Slovic suggests, "By confronting face-to-face the separate realm of nature, by becoming aware of its otherness, the writer implicitly becomes more deeply aware of his or her own dimensions, limitations of form and understanding, and processes of grappling with the unknown" (4). Therefore, what develops from these texts is an invitation to the audience to consider its own relationship to nature and to acknowledge the limits of human comprehension about the natural world—a move that certainly carries the possibility of moving one toward a more ecocentric perspective.

This confrontation with nature and the heightening awareness that emerges defines much of the early nature writing about the desert. Take, for example, Mary Austin's *The Land of Little Rain* (1903), one of the most famous examples of desert writing, which chronicles her years on the eastern ramparts of the Sierra Nevada.[7] Attuning her senses to the unique characteristics of this portion of the Great Basin, Austin discovers the misaligned perspectives that she and others possess when considering nature. In her chapter "Water Trails of the Ceriso," Austin describes the faint paths that small animals have made across the land in search of water. She notes, however, that such trails are hardly evident to the casual observer and that only by "getting down to the eye level of rat and squirrel kind, one perceives what might easily be wide and winding roads to us if they occurred in thick plantations of trees three times the height of a man." By reorienting her view to see the land from an animal's perspective, Austin finds a new world opening to her. At the same time, she realizes the limitations of human perception as she laments, "It seems that man-height is the least fortunate of all heights from which to study trails" (11). This revelation invites a thoughtful reconsideration of the human relationship to nature, and, more specifically, to these water trails through arid country, as it expresses the belief that our current understanding of the desert may be flawed. In the decades following Austin's work, numerous writers would follow her lead as they attempted to reorient the audience's consideration of the Colorado River watershed's deserts by championing their virtues while also creating room for their respective voices to emerge.

Dialoguing with the Desert—Contemporary Desert Literature

Up to this point in my discussion about natural history and the nature-writing traditions, I have assumed a broad understanding of "nature." However, from the advent of the American nature-writing tradition, a much narrower vision of the natural world ensued. Because the genealogy of this genre goes back to the well-watered and -cultivated lands of England and America's eastern coast, nature writers promoted the pastoral landscape, one characterized by "the quiet peace of hay barns, orchards, and mountain valleys" (Worster, *Nature's Economy* 16). Not until the late nineteenth and early twentieth centuries did deserts take on an equally celebratory role as perceptions about the American West and the Great American Desert began to change through the frontier, garden, and wilderness myths to create what I view as a desert pastoral. And it is primarily within the deserts of the Colorado River watershed where the American desert literary tradition materializes. Like the vanguard natural history and nature writers who turned to the villages, farms, and forests around them, writers today have turned their attention to the region's sandstone canyons, alkali basins, and saguaro-studded hills to come to a greater understanding of their processes and of the writers' own reactions to these desert lands. Their observations constitute a reckoning with landscapes equally puzzling as those the explorer-scientists of previous centuries encountered and necessitate a broader range of discourse by which to account for such unconformities.

Perhaps no desertscape within the watershed has caused more confusion, bewilderment, and awe than the Grand Canyon. Stretching nearly three hundred miles through northern Arizona, it comprises portions of the Great Basin and the Mojave and Sonoran Deserts. For newcomers to the canyon's inner depths, the abundance and variety of life, the exquisite geological features, and the river's ever-changing form can quickly leave one lacking for words, most likely as Cárdenas was when he first peered into its depths. For a seasoned Grand Canyon traveler like Ann Zwinger, the effect is consistently similar. Remembering a remarkable two-week Colorado River trip through the Grand Canyon, Zwinger writes in *Downcanyon* (1995) of her long hike to the rim and the reentrance into the busy, civilized world. Exhausted and disheveled from weeks of hiking and backcountry living, she encounters a tourist in a parking lot who asks Zwinger, "Is there anything down there?" Caught off guard by this woman's question, Zwinger provides a brief, unmemorable response, which supposedly seems to have satisfied the inquisitor. However, this innocent yet inane question to anyone who has ventured inside the Grand Canyon "haunted" Zwinger for some time. Only when later flying high over the river and canyon did Zwinger happen upon a satisfactory answer. She explains, "There *is* something down there, and it cannot be explained in a listing of its parts. It has to do with truth and beauty and love of this earth, the artifacts of a lifetime, and the descant of a canyon wren at dawn" (237). Zwinger further clarifies this response, suggesting that what one finds in the canyon is a nearly indescribable blend of large and small, life and death, intimate and distant. Thus, a great many things exist "down there," yet they defy easy summation. Ultimately, Zwinger settles upon this simple idea: "the terrible life-dependent clarity of one atom of oxygen hooked to two of hydrogen that ties us as humans to the only world we know" (238). Zwinger's initial bewilderment at defining this particular portion of the watershed and her ongoing difficulty at adequately describing what is really down there have the author grappling for a number of different images and ways of knowing that articulate her observations. From such subjective and abstract concepts as "truth" and "beauty" she reduces her personal response and the emotions embedded in it to the very tangible and measurable reality of water's elemental composition.

This vacillation, between the subjective and the objective, the conceptual and the concrete, the poetic and the scientific, is a pervasive commonplace of desert writing evoked time and again as explorers, scientists, and artists have struggled to articulate the desert before their eyes. So often, the writers return to paradox and the desert's uncanny ability to resist easy classification and categorization.

To address this tension, many writers represent what they perceive as the desert's agency. As Catrin Gersdorf argues in *The Poetics and Politics of the Desert* (2010), the desert is "an entity that *does* something rather than *being* or *meaning* something" (1). Reading the desert as an agent casts the desert in a significantly different light than the medium upon which humans project their emotions and desires, as Limerick and Wild suggest in this chapter's introduction. While the early-twentieth-century writings continue to position deserts in this manner, more recent literature attends to what many writers view as the desert's agency. Situated squarely within the natural history tradition of close observation of nature's flora and fauna, these writings turn to the deserts to cast a focused gaze on their biologic and nonbiologic processes, from which the writers argue that the desert does indeed do something rather than just operate as a template upon which humans project their desires to be easily quantified and analyzed. Indeed, their engagement with the desert counters traditional attitudes toward nature represented in so many of the explorers' narratives that all too often have reduced the desert to a resource for exploitation or a place to avoid.

Take Joseph Wood Krutch's close observations in the Sonoran Desert, for example, which lead to his lessons in desert living that emphasize the need for space and a new type of habitation. In his chapter "What the Desert Is Good For" from his classic *The Desert Year* (1952), as well as in his later book *The Voice of the Desert* (1954), Krutch addresses those attitudes toward the desert that consider its value and show how current views have led to exploitation through nuclear testing. While the desert for many is valuable only because, ironically, it is apparently worthless—therefore, the perfect place to detonate bombs—Krutch's careful observations of the desert posit a different valuation of this place. His awareness of his surroundings teaches him about the importance of sparseness. Reacting to the "Malthusian dilemma" he sees plaguing the earth in both human and nonhuman communities, he looks to the desert for answers. While he acknowledges that overspeciation probably exists in the desert, he writes,

"I cannot help seeing as an advantage the simple fact that the land here is dry enough to prevent uncomfortable crowding, and I cannot help wondering if one of the worst features of most of the world in which we live is not the simple fact that, to an ever increasing degree, mere living space is the thing which gives out first" (*Desert Year* 88–89). Fearing the loss of room for humans to inhabit, Krutch draws from the sparseness of desert flora and fauna imposed by aridity to demonstrate how each species not only survives but flourishes.

Later in the text, Krutch becomes more enthusiastic about the desert's ability to point the direction toward better living. Having returned from a trip that took him out of the Sonoran Desert and into the Mojave and Colorado Deserts of Southern California, Krutch reflects on what he observed on this journey and concludes that the desert is one powerful teacher: "Call it, if you must, only another aspect of the pathetic fallacy, but the desert seems to approve and to encourage an attitude with which I have found scant sympathy among men, and of which I have never before been quite so sure that even nature approved. However fanciful this may seem or, for all I care, however fanciful it may actually be, all the deserts seem to suggest and confirm a system of values for which much ought to be, but very seldom is, said" (180). Reacting to human disregard for the desert and what it might teach to the attentive observer, Krutch creates a work attesting to the desert's power in communicating these values and transforming his perception of the place. Admitting how he has learned from watching the desert, he moves into a discussion regarding man's insatiable desire for material possessions and contrasts human behavior with the desert's processes. Of human society he contends, "The all but universal ambition of the individual and the all but invariable aim of every proposed social or political movement is to get, for oneself or for others, more *things*" (80–81). He then juxtaposes Western society's insatiable wants with the "economy of scarcity" that desert life embodies (*Voice of the Desert* 99). Observing his surroundings, he writes, "The very fauna and flora proclaim that one can have a great deal of certain things while having very little of others; that one kind of scarcity is compatible with, perhaps even a necessary condition of another kind of plenty—for instance . . . that plenty of light and plenty of space may go with a scarcity of water" (*Desert Year* 181–82). His observations of the kangaroo rat—"the mouse that never drinks"—stand as a powerful reminder of how one animal has adapted and thrived under conditions that most people would find impossible.

So it is with many of the other organisms Krutch examines in his time in the Sonoran Desert. Of the spadefoot toad that lies dormant during much of the hottest and driest parts of the year, he suggests it "represents, in a very extreme form, one of the ways of living in the desert—namely by lying low most of the time" (*Voice of the Desert* 30). The roadrunner, on the other hand, stands as a symbol for those newcomers to the desert that have adapted to its conditions rather than impose their own. He notes how the bird's "peculiarities represent things learned, and learned rather recently as a biologist understands 'recent'" (38). Despite the roadrunner's adaptability, Krutch acknowledges how current attitudes seem to disparage the bird. He writes, "It seems that a creature who so triumphantly demonstrates how to live in the desert ought to be regarded with sympathetic interest by those who are trying to do the same thing" (38–39). The problem exists, however, that many people do not take time to learn from this animal and what it could teach about adapting to a new environment. Krutch believes that although the roadrunner and the quail are enemies, they continue to coexist as "neither seems likely to eliminate the other." But as he points out, "Man, on the other hand, may very easily eliminate both. It is the kind of thing he is best at" (39).

Recognizing humans' inability to adapt sufficiently to the Southwest's climate and geography, and their eagerness to control what they do not understand, Krutch concludes both texts with a charge to humankind to rethink its impact on the desert. He reminds the audience that a balance between abundance and scarcity is necessary to desert living and that "to have experienced [the desert] is to be prepared to see other landscapes with new eyes and to participate with a fresh understanding in the life of other natural communities" (223). Along with this call to see the desert anew, Krutch argues that ultimately people need to love the desert in order to save it. Looking to Aldo Leopold's famous essay "The Land Ethic" as a model for approaching the desert with a fresh perspective, Krutch concludes vis-à-vis Leopold that what lacks in current interaction with the desert "is love, some feeling for, as well as some understanding of, the inclusive community of rocks and soils, plants and animals, of which we are a part" (193). Throughout his texts, Krutch works toward understanding the unique desert community of which he sees himself a part and urges his readers to learn from the desert and to love it in order to save this unique ecosystem. He seeks to overcome the "sentiment and science" divide that marks much of the nature writing of the early twentieth century and the nature-culture binary that has dictated much of the literature about the natural world (Lutts 3). From this synthesis of discourses emerges a refashioning of the arcadian model of natural history along with all of its idealism. Krutch adapts the formula to create a desert pastoral, one that Abbey adopts in *Desert Solitaire* as he structures his own narrative on seasonal cycles, inspects his "gardens," tends to the surrounding flora and fauna, and basks in the space and relaxation such locales afford.[8] Through his close, quasi-scientific observations that then give way to reflections that celebrate the flora and fauna and advocate sympathy and love by humans for the natural world, Krutch directly calls for a more sustainable and harmonious community between humans and nature.

Far from the rolling green hills of western Europe's fertile valleys, Krutch suggests that peace and repose denote his home country. Like the traditional pastoral, which positions itself as a return to simple living much like the wilderness myth does, Krutch's desert pastoral also advocates this view. However, while the traditional pastoral represents a flight from reality, Krutch's version embraces a modicum of reality rather than rejecting it for a more ideal setting. Recognizing actual environmental conditions and limitations, Krutch's pastoral vision suggests that the panacea to modern living is found not in the literary imagination of the pastoral but in adapting to one's surroundings, taking note of the lessons nature can teach those willing to listen. As Krutch employs the pastoral to describe the desert, the reader could expect his presentation of the ecosystem to follow a similar trajectory. Yet by foregrounding the knowledge he gleans through his astute observations and paradox, which complicates this motif, his texts become far more insightful and relevant as they posit more sensible ways of living in the Sonoran Desert.

The Desert on Its Own Terms

While the desert pastoral becomes a staple in the desert writing of the mid-twentieth century, writers like Childs move away from this motif and the outright celebratory tone that Krutch and others bring to the genre. Similarly, Childs brings a more trained eye for objectivity that dissuades him from making more overt, didactic statements about what the desert should teach humankind. Make no mistake, however; Childs certainly has a soft spot in his heart for the deserts of the Colorado River watershed that lead him through the mountains and canyons of Arizona and southern Utah. But his admiration and interest derive from a different background than most of the modern desert writers. While many of the great chroniclers of the desert and the larger American West, for that matter, like Charles Fletcher Lummis, John Van Dyke, Austin, Krutch, and Abbey, are all midwestern and eastern transplants to the Southwest, Childs was born and raised within the West's arid regions. As he opens *The Secret Knowledge of Water*, he is very much aware of his well-established desert roots.

He begins the book by emphasizing his long connection and experience with desert water, mentioning the spring by his mother's birthplace that nourished her and eventually him as her offspring as well as early childhood memories of southern Arizona that include a cottonwood tree, tadpoles in a dwindling puddle, and a small desert stream coursing over rock (xii). Beginning the text with these images establishes his authority as a keen observer and a native to the desert, one whose observations are supposedly more trustworthy than those outsiders who have retreated to the desert with a nostalgic longing for a bygone era of pristine wilderness or a more salubrious setting. Reflecting on an experience within one of the desert's countless canyons, Childs admits that "I realized that part of my life was here, something I would have to seek with full attention, dictated by the water from my mother's spring sent from her body into mine" (xvi). As he further emphasizes his connection to the desert as home rather than desert as destination, Childs's treatment of desert water in these opening pages reflects Jane Bennett's "thing-power," "the curious ability of inanimate things to animate, to act, to produce effects dramatic and subtle" (6). From this opening reflection to the experiences with water throughout the text, Childs homes in on desert water's agency and its ability to influence his life though its effect on both his physiology and his creativity as a writer. To understand water's influence in shaping all that comes in contact with it, Childs appeals to scientific discourse to provide a more objective and possibly more accurate vision of the desert and its water than one steeped in a Romantically charged imagination.

Nevertheless, as Childs's comment above suggests, his quest is just as much a journey into the self as it is one in pursuit of desert water. He seeks that part of his life connected to the desert's water and explores the genealogy of that water which courses through his veins. Therefore, for Childs to understand these entities, he must understand himself. For him, truth is at the heart of his quest into the desert and into his soul. Unlike a quest for scientific veracity apparent in the passages above, this desire to know the truth of his life is more poetically motivated as it emphasizes reflection. Thus, from the text's outset, different perspectives and means of expressing them arise in the author's epistemological and ontological journey. Ironically, as Childs's introductory comments strive to establish his connection to the desert as an insider, he nonetheless adopts the clichéd motif of entering the deserts to find truth, a theme stretching back thousands of years as people retreated to the desert for enlightenment and that flourishes today through the wilderness myth that promises its followers a more authentic, truthful existence among nature's wild places. As Wild observes of this acute attentiveness to pursing knowledge in the wild, "the focus becomes so concentrated that the desert—the supposed subject—begins disappearing, replaced by the viewer's own reflected image. It is a journey from expansiveness into self-absorption" (*Desert Literature* 11).

However, Childs's writing resists a full-blown paean to the self. Instead, he shifts from a subjective to a more objective position as he defers to his scientific training acquired through graduate studies in desert ecology at Arizona's Prescott College to bolster his position. Raised between homes in Arizona and Colorado, Childs spent much of his youth exploring the deserts and rivers of the American Southwest on foot and by boat, becoming intimately acquainted with the unique characteristics of this region. Underlying much of Childs's writing, and what is clearly visible in *The Secret Knowledge of Water*, is his attention to and reliance on the scientific method as an organizing principle. Chasing a burning question about the desert, Childs enters it to test hypotheses, record observations, and work toward conclusions. Consider his remarks in the closing chapter of this text, which speak to his methodology: "I had spent two years tracing the bloodlines, meticulously studying the documents, then walking to see if it was true, if the desert was, indeed, bound by water as I had believed" (266). While this statement reveals the overall rationale for his travels, it also suggests a reliance on history, tradition, and procedural knowledge to inform his scientifically motivated excursion. Rather than create an ahistorical desert where he will make first tracks, Childs is keenly aware of the long human presence within the desert. Thus, as he embarks on his various treks, he often follows the paths others have blazed before him, willing to acknowledge their contributions to the region's body of knowledge and his own gaps in understanding.

Pursuing a question and a hypothesis, Childs finds himself in some of the most remote locales of the continental United States where his preparation in desert ecology helps him reconcile the anomaly of desert water. In the text's first section, "Ephemeral Water," Childs travels to Thousand Wells along the Arizona-Utah border to search for water pockets in a land of pure rock and the Cabeza Prieta National Wildlife Refuge in southern Arizona's Sonoran Desert to map water holes, or *tinajas*, as part of the refuge's commitment to preserving and protecting desert bighorn sheep. In the case of the Thousand Wells region, which John Wesley Powell once described as an area of "innumerable basins" (*The Exploration* 333), Childs seeks out possible water pockets after coming across one man's graduate fieldwork about a tadpole shrimp. This small desert crustacean, referred to by Childs as *Triops*, happens to be "the oldest living animal on the planet" and requires a reliable pool in which to propagate (*Secret Knowledge* 47).[9] Taking a page from Krutch's work, Childs draws wisdom from *Triops*'s life cycle to understand more clearly human limitations and possibilities in the desert.

Through Childs's consideration of *Triops*'s relationship to ephemeral water pockets, the author negotiates various epistemological and ontological systems and explores the human-nonhuman interface. Childs considers the survival mechanisms of this species, which must endure extreme drought, and informs the reader of the process of *anhydrobiosis* or, more simply, "*life without water*." As the author explains, this is a condition in which the shrimp "shrivel up until they are dry as cotton balls, releasing all of their water" (61). From this basic description, Childs goes into a multipage explanation about science's understanding of these organisms. He refers to their ability to produce trehalose, a special compound to enter this particular state, and their ability to withstand X-rays and radiation without harm, go decades without water, and endure constant temperature fluctuations (61–62). He also describes how this species, along with numerous other organisms, lives through *phenotypic plasticity*, or "the ability to alter the body's shape in step with its environment" (67). Thus, as the pool shrinks, *Triops* slows its biologic processes down; as the pocket grows with additional runoff, *Triops* grows. Through this point in the narrative, Childs's text is motivated scientifically as he seeks to understand the phenomenal behavior these organisms exhibit.

At the same time, however, the adaptations that Childs describes become a larger metaphor for the chapter "Water That Waits" and thereby seem to be aesthetically or poetically motivated as he sets up the metaphor to help the reader draw connections between *Triops* and humans. Beginning the chapter by recognizing his own limitations in the desert and the amount of water he has to carry in order to survive just a few days, he turns his attention to what he calls the "prophetic knowledge" of *Triops* and other similar organisms as they easily adapt to their surroundings. From his observations of this and other desert creatures, Childs concludes, "My own life had to be measured by completely different standards" (69). The juxtaposition he creates between his own existence and that of *Triops* through his scientific inquiry and research allows him to marvel at the resilience and permanence of this creature and also recognize his and society's limitations in attempting to adapt to the harsh conditions imposed by aridity.

Childs's attention to this unique organism and its ability to thrive in conditions that would quickly kill a human being evokes a Krutch-like didacticism that defines much American desert writing. True, there are lessons to glean from Childs's consideration of *Triops*. We can certainly read his interest and dedication in understanding this species' survival as an admonishment to try to adapt in our own way, to not become habituated to more water than is necessary, to work with what is available, and to make the most of the circumstances. We could even extend this vignette's line of reasoning to suggest that if humans could just learn to adapt to the unpredictable nature of water, then perhaps, in the course of time, humans could evolve to pattern their own biological processes to better align with desert water's ephemerality. Yet Childs never comes to these conclusions, at least not explicitly. He never turns away from the dominant scientific narrative to suggest to the reader "and this is why this organism is important for us to know about." Rather, he lets the science speak for itself as he describes *Triops*'s qualities. The most we gain from Childs is his recognition that he and his traveling companion are worlds apart from *Triops* in their ability to live in these harsh conditions. "Tom and I were lithe, short-lived creatures who would never know how to sleep for an entire hundred years," Childs reflects. "This was a different strand of life from my own" (69). But by deferring to science to describe *Triops* and its incredible adaptations, Childs paradoxically grants nature a voice through the persistent metaphor that lurks beneath the surface of the objectivity his writing evinces.

The impacts of this choice to foreground science are manifold. By emphasizing *Triops*'s physiological properties rather than digressing into his connection to the species, Childs allows the audience to come to its own conclusions about this creature and its relationship to the desert. Likewise, by invoking the organism's otherness—its existence in a world far different from his own—Childs prevents himself from purposefully using *Triops* as a metaphor in the service for human society. Instead, he foregrounds scientific discourse and thereby exercises, surprisingly, a sense of restraint and distance that derives from what objective and accurate scientific inquiry demands. In establishing the known facts about *Triops* and then resisting the desire to further an agenda that comes from these observations, even if a benign one, Childs creates a nuanced vision of the desert. While reliant on a system that seeks to explain all the complexities of natural phenomena, he nonetheless creates an alternative portrayal of nature that enables it to exert its own voice—a true ecocentric gesture that grants him an even greater ethos as an expert explorer and observer of the desert.

Although Childs builds his narrative on the scientific method by formulating hypotheses, developing methodologies, and then testing results, he recognizes its inability to capture the full mystery of the region and looks instead to other ways of knowing to aid his growing comprehension. In his 2004 book *The Way Out*, Childs and his hiking partner find themselves in a remote corner of the Colorado Plateau, struggling to negotiate a path through an expanse of cryptobiotic soil. Their hesitancy to move forward derives from their understanding of this crucial component of healthy desert ecosystems, this living soil composed of cyanobacteria, lichens, and mosses that provides nutrients and stability for other organisms (Belnap). Trudging ahead with no alternative route across, Childs considers the implications of his actions:

> Science is easy for me, with all of its neatly turning gears and ratios. I can put my weight against it in discussions with learners of obscure disciplines: fluvial geomorphology, osteology, microbiology. But I have also known that I should not put my weight there. With this limited human spectrum of sensations, our blithely unquestioned bias of self-superiority, how can I possibly confide in the imagined purity of science? Devastated beneath my boots are colonies of rare desert mosses and these brittle, creeping colonies of blue lichen. Spore heads no larger than drops of mist bind and crush. Hard science barely allows for mystery, for true sacrifice and loss. It robs this soil of what I see right now. (*Way Out* 87)

This brief reflection amid Childs's larger purpose of relating how he and his friend attempt to navigate their way through a maze of sandstone speaks to Childs's ongoing negotiation of the role of science and other epistemologies in his work as he seeks to understand the deserts of the American Southwest. It also underscores his awareness of the larger human ecological footprint upon the earth and the materiality of the microbes with which he interacts. Evident in this passage is Childs's "attentiveness to matter," which Bennett believes "can inspire a greater sense of the extent to which all bodies are kin in the sense of inextricably enmeshed in a dense network of relations. And in a knotted world of vibrant matter, to harm one section of the web may very well be to harm oneself" (13). Childs's understanding of the life processes of cryptobiotic soil, gained through close observation and his scientific training, as well as his attentiveness to wonder and this complex entity composed of organic and inorganic matter allow him to see this living dirt as a key player in healthy desert ecosystems that influence human systems. Childs speaks of stepping on this special soil as "sacrifice and loss," as footprints kill the organisms that mitigate erosion and transportation of desert soils.[10]

Similar to his observations in *The Secret Knowledge of Water*, this passage avoids didacticism and opts instead to have the science and the image he creates speak for themselves to offer possible correctives to how humans interact with the natural world. However, while his technical expertise leads him to understand the role of this living soil and see its connections to the larger ecosystem, it nonetheless fails to adequately convey the effects of destroying one of the oldest organisms on the planet. Placing limits on scientific knowledge, Childs opts for other epistemologies to engage his surroundings and recognizes that scientific knowledge is inadequate and incomplete in his quest to understand the desert. As a result, Childs argues for an alternative approach to the desert that evades the promises of his professional training in favor of a more poetically inflected discourse that further allows the desert's voice and agency to emerge.

Childs begins to question science's adequacy in conveying the truths of his observations years earlier as depicted in *The Secret Knowledge of Water*'s first chapter, "Maps of Water Holes," and emphasizes this realization in much of the remainder of the text. Throughout, Childs finds himself teetering between the utility of empirical fact and science's limitations to provide a complete picture of the desert. "When I began hunting for water in small, isolated holes," Childs explains, "I had wished for a tangible knowledge, a line of information I could personally own" (*Secret Knowledge* 75). He continues in great detail as he includes a typical entry of his measurements:

> I had taken notes carefully and made my own maps, quantifying whatever I could measure. *On the top of the Navajo sandstone at the end of the summer rains. In steep, narrow canyons of granite, not in the arroyos below*. To prove my knowledge, I wrote academic papers on the positioning of waterholes, spatial distribution, ways of determining longevity, submitting these to scholarly reviewers for a master's degree program I had applied to. My measurements of water holes in Cabeza Prieta came out in stunning, colored charts. Weeks of fieldwork from the wildlife refuge appeared in fifteen pages of:
> Pool #33 104 1 (28 gallons)
> *Coordinates*: N 32° 20' 15.1" W 113°48' 36.4"
> *Elevation*: 1,650 ft
> *Depth (h)* = .18m
> *Average width (2r)* = 1.5m
> *Protection*: Protected
> *Local Catchment Area*: Large
> *Sediment*: Low
> Turbid
> *Invertebrates*: Mosquito larvae/pupae, chironomid larvae. (75–76)

As in his commentary on *Triops*, this broader explanation of his research about water holes clearly demonstrates Childs's reliance on his scientific training to make sense of his observations and experiences. But his time in the Cabeza Prieta quickly teaches him that scientific knowledge alone cannot reveal the secret knowledge of desert water. As Childs's text progresses, he becomes more and more aware that the knowledge he believes to possess about water is insufficient and unsatisfactory. Whereas water holes are, in his words, "effortless to study [because] they have discrete boundaries that take easily to tape measure, a global positioning device, or a Brunton compass," moving water, the subject of the text's next section, is, in the author's words, "different" (76). Because moving water "furrows itself into shapes as it runs, immediately telling stories out loud," he suggests that "an alphabetic string of symbols is left in sand and on rock faces after it passes." While trying to decipher these marks, Childs concludes that the knowledge he had accumulated about the water holes "turned suddenly arcane and restrictive. The knowledge was no longer so simple to possess. It was not as innocent as *where* and *how much*. It was now asking questions of me" (77). At this point in the narrative, Childs confides that this training grasps insufficiently the phenomena of desert water that defy his understanding, a recognition that suggests the desert is far more than a blank slate on which to project one's own agenda. Despite Childs's efforts to this end, he discovers that desert water resists these attempts, as it behaves outside of his ability to control and order it according to his liking and technical expertise. In this case, the desert exerts its own agency and speaks to Childs, teaching him that its water is far more complex than his training suggests.

To make better sense of his observations, Childs changes course to pursue another means of understanding not typically part of the objective scientist's program: storytelling by those who have come before him to this parched land and who have an intimate knowledge of its unique features. He writes, "Familiarity with scattered water holes has become obsolete, left only for the bighorn sheep. Words are now missing from the story of ephemeral waters, severing critical pieces of information. Many people have died while crossing this desert.... They died because the story was forgotten." Childs relates a powerful lesson here for desert dwellers, suggesting that our present society has forgotten the stories and instead tries to engineer its way out of reality with its "cement aqueducts to siphon distant rivers, and holes...drilled into ten-thousand-year-old groundwater" (10). Yet, again, he skirts overt didacticism and blame of any particular group who today has forgotten what kind of relationship it takes to live sustainably in the desert. Instead, Childs makes his observation very matter-of-factly, leaving the audience to consider where it stands in relationship to water, the desert, and living with aridity.

In order "to put a story back together and recover parts that had been lost" (9), as Childs suggests, he embarks on a quest to consult the documents of previous explorers and to read the land more closely to see what information it can yield. As a contemporary desert traveler, Childs continues the tradition of his predecessors to the Southwest's deserts who repeatedly sought time and again to restore the knowledge that had been lost or forgotten by previous expeditions. Turning to the travels of Father Kino, Childs defers his own knowledge to an earlier voice, relying on centuries-old knowledge to move from one water hole to the next. Nevertheless,

Childs does not rely solely on Kino's experiences to find the water; he relies on much older knowledge inscribed on the land itself. Traveling back and forth between a number of known *tinajas* that allow him to extend his search deeper into the refuge's mountains and valleys, Childs notices faint lines or trails that crisscross the desert floor in logical patterns. Reminiscent of Austin's commentary on animal pathways to water, Childs's "waterlines" are the ancient paths of the Indigenous people who knew where to find reliable water as they migrated through the area. What is most significant about these waterlines is that they are, as Childs insists, "the opposite of canals, moving people to water rather than water to people." This new realization about how previous generations survived in a challenging environment reinforces the author's previous observation about how our culture has forgotten the stories. In each case, Childs's comments imply alternative means by which to engage the desert: "Know the land and its maps, you might live" (31). Likewise, he suggests that society's current process of bringing water to cities, often hundreds of miles from the source, is out of step with a deeper, more sustainable knowledge that earlier generations possessed. The paradox here exists in our reluctance to see the old knowledge inscribed on the land that speaks to the environment's limitations. Despite our sense of progress and achievement, Childs's comments suggest that in order to survive, a "new" paradigm of success actually lies within our ability to understand and embrace lessons from the past and unlikely sources.

It becomes clear the further one proceeds through the text that Childs cannot achieve a complete knowledge of desert water through science alone. Whereas the discussion about *Triops* and water holes grounds the text in view of a scientifically constructed reality, the book's third and final section makes abundantly clear that Childs recognizes that knowledge comes in various forms, including from nature itself. As "Fierce Water" describes the floods that leave death and destruction in their wake and pushes Childs to make his own life-and-death decisions, his text also reveals his negotiation between scientific and poetically attuned discourses as he searches for answers to events he struggles to understand. Where he could once analyze and measure water in numbers and figures, the floods come so unpredictably to the desert that they leave him longing for a means of quantifying their behavior. In the end, he concludes that "it was no longer my own longing or my own body, not some piece of knowledge I could possess. Water now had the knowledge" (167–68).

While science punctuates this last section to account for a devastating flood preserved in the Tohono O'odham oral tradition, explain terms like "hydrologic jump," and outline a university researcher's work on canyon erosion, Childs foregrounds story throughout this third section and its relationship to scientific knowledge to explain the paradoxical nature of these floods (240). In the case of the Tohono O'odham account, Childs teeters back and forth between the knowledge that science and Indigenous storytelling reveal. According to the narrative, a hunter once pursued a badger into a hole, and after poking his stick into its dark recesses, the hole exploded in a torrent of water that swept away a number of villages and threatened to drown all the people. Having exhausted all other options of stopping this sudden burst of water onto the desert's sands, the people decided to sacrifice four children to the flood, throwing them into the hole with hopes of stopping the flow and saving the community. To their relief, the sacrifice proved sufficient, and, as the water receded, a large rock fell into place, capping the hole (176–77). At this site, the people erected a shrine to the children, and it is this gathering of rocks and other paraphernalia for which Childs searches. Of this quest, he explains, "I was scientific about it," as his keen eyes recognize "an embankment of compacted, water-driven sand left far from any drainage, just out in the desert" (177).

At the same time that he is trying to be "scientific" in his search, he also turns away from a scientific explanation of this event. "Purposely," he admits, "I did not mull over records of local geohydrology to isolate this story of water bursting out of the ground" (182). Instead, Childs chooses to emphasize the Indigenous account. While he juxtaposes the narration of these events with information about general groundwater levels throughout the Southwest, Childs opts for this version of the water's sudden appearance, as it teaches him an important lesson about water's relationship to the desert. "Our offerings to water, our requests of it in the desert, must be balanced carefully. Not too much and not too little," he suggests (183). While this statement argues for a tempered desire for water, as well as a more comprehensive awareness of the Indigenous knowledges shaping the cultural and physical landscapes (a point taken up in greater detail in chapter 5), it also mirrors Childs's balanced approach in his consideration of the offerings made to water. They cannot be solely entrenched in science but should also involve story and poetic discourse to provide a broader spectrum of what the reality of desert water entails. Such an approach to understanding water differentiates Childs's work from earlier explorers who place the desert's worth in the realm of science's ability to dictate value. This negotiation also separates his writing from more contemporary nature writers, who draw on the natural history tradition and often make more didactic observations about human relationships to the natural world. Rather, Childs acknowledges that science has clear limitations, and its ability to procure and endow understanding goes only so far. Similarly, he refrains from making extended commentary about how we should live with the desert and its water, choosing instead to let the desert's voice, as captured through paradox and the gaps between scientific and poetic discourse, do the talking. In the case of "The Sacrifice of Children," Childs defers his technical training to give space to another perspective on desert water, thereby placing Indigenous ways of knowing on similar footing with more historically dominant epistemologies.

As the rest of the section unfolds, Childs's understanding of water comes less and less from his academic knowledge and increasingly from other sources that include the stories to which he is willing to listen. In this way, his writing becomes, as Teorey suggests, "a two-way conversation that requires humans to interact with Nature on its terrain" (11). Childs tells, for example, of the three hikers traveling up Phantom Creek in the Grand Canyon, where a husband and wife drown while the brother-in-law is somehow able to ride on top of the flood and survive (*Secret Knowledge* 216). Likewise, he considers the fate of twelve hikers caught in a flash flood in Antelope Canyon, a narrow slot on the Navajo Nation. Only the guide survives, having been entirely stripped of his clothing and left naked on a ledge by the raging waters (217). And when he recalls finding himself caught in a canyon with a massive flood bearing down on him, he confesses, "I had been studying water. I had read hundreds of scientific journal articles, taken innumerable pages of notes, produced papers, articles, treatises on the performance of water in the desert. It was all washed blank here" (272). Childs echoes these exact sentiments in his larger work on flash floods, *The Desert Cries* (2002). Herein he admits, "Even all of my studies appear foolish as I try to frame the flood, to break it down into math" (138).[11] Like many of the early explorers who ventured into the desert unknowns armed with the knowledge of cartography and other developing fields, Childs initially expects that scientific inquiry will provide the most reliable means of quantifying the desert's unique characteristics. Yet despite his extensive technical training in the complex and paradoxical phenomena of desert water, he eventually finds this knowledge insufficient to fully explain the hows and whys of this mysterious entity. Paradox thus becomes the prominent theme in his writing as he recognizes, as have the explorers and numerous desert writers before him, that this rhetorical device is the best way to articulate the irony and wonder these individuals attribute to the desert.

The Voice of the Desert

The question still at hand is what effect does Childs's ongoing negotiation with scientific and other forms of knowledge that come through Indigenous stories and anecdotal evidence have on the larger understanding of the American Southwest and the Colorado River Basin in particular? As a twenty-first-century explorer with roots sunk deep into the explorer-scientist tradition of the nineteenth century, Childs's work provides one approach to understanding the spectrum of knowledge possible when engaging the watershed's deserts. By employing the well-used trope of paradox in writing about the desert Southwest, Childs brings a number of seemingly contradictory perspectives into play. With his academic training in tow, Childs aligns himself with nineteenth-century explorers who relied on objective observation, precise methods, and thorough experimentation to break down nature's complex phenomena. At the same time, however, Childs allows for vestiges of the Romantic nature-writing tradition and Indigenous oral traditions to infiltrate his work as he creates room for a more mysterious, potentially unknowable desert that resists empirical science's reach.

As Childs intermingles scientific and poetic discourses, he invites nature's own knowledge and voice to surface. While critics have scorned the pathetic fallacy and anthropomorphism as literary liabilities, they nonetheless perform a very powerful function in nature writing. Lawrence Buell dubs the pathetic fallacy "nature's personhood" and argues that to prohibit its use "would be worse than to permit its unavoidable excesses. For without it, environmental care might not find its voice. For some, it might not even come into being" (*Environmental Imagination* 218). It is through this tool that an author weds "ecology to ethics," thereby creating a useful technique by which to influence an audience's perspective on the natural world (201). This approach then becomes what Bryan L. Moore calls "ecocentric personification," a further development of the pathetic fallacy or nature's personhood, to "*persuade* an audience that all living things are *connected*" (10). This iteration of the pathetic fallacy aligns closely with the "psychological phenomenon of 'awareness'" that likewise functions to provoke readers to consider the natural world, and the desert in Childs's case, on a more profound and personal level (Slovic, *Seeking* 3). At the same time, they allow nature to have a voice of its own—not literally, of course, as the words we read are the author's, but by a humble gesture of acquiescence and deferral to nature's own processes.

More recently, Jane Bennett has argued that anthropomorphizing nature (or matter) might also have merit in establishing a more ethical relationship to nature. "Maybe a bit of anthropomorphizing will prove valuable," she posits, "maybe it is worth running the risks associated with anthropomorphizing (superstition, the divinization of nature, romanticism) because it, oddly enough, works against anthropocentrism: a chord is struck between person and thing, and I am no longer above or outside a non-human 'environment'" (120). Viewing the nonhuman in this regard, as a fellow sojourner on earth, "tends to horizontalize the relations between humans, biota, and abiota. It draws attention sideways... toward a greater appreciation of the complex entanglements of humans and nonhumans" (112). Thus, while *The Secret Knowledge of Water* primarily employs scientific and poetic discourse to understand nature, this attention to evoking nature's voice, which then surfaces as an instructor and a behavioral corrective, creates a fusion of three different epistemological positions that continuously mingle in light of the various circumstances in which the author finds himself. Together they become emblematic of what Teorey sees: "The respectful collaboration of nature's voice and human language [that] can reestablish a story of health and prosperity" within these bioregions (7). This story constructs a new reality, one based on the existence of the desert's agency, autonomy, and own knowledge that Childs reveals through his close engagement, patience, and willingness to recognize and accept his otherness from the desert. But rather than use this difference as a way to exploit and profit from the desert, it invites a reverence for a closer engagement with an entity that exceeds humans' ability to fully comprehend its mysteries.

Childs's deferral to the desert's voice through his use of paradox relies on harmonizing divergent epistemologies into a greater whole. As he does this, Childs resurrects nineteenth-century scientific discourse and its various branches to create a new vision of the Southwest's deserts. To achieve this, Childs synthesizes the disciplinary disconnect Laura Dassow Walls views between science and the humanities: "Insofar as each discipline is founded on a single mode of vision, then no one discipline can cash out the entire universe: not literature for all its scope and beauty; not science, for all its range and power" (200). Childs's text embraces this perspective as he comes to explain that science is not the definitive source of knowledge, while story, or literature, is not entirely adequate on its own, as it is too easily altered or forgotten over time. Along these lines, we can read *The Secret Knowledge of Water* as an early manifestation of the environmental humanities and their attempts to synthesize various disciplines and their respective methodologies to better address current environmental crises.

Viewing Childs's text in this light positions it as a contemporary embodiment of a more integrated and interdisciplinary approach to knowledge that Walls observes in the work of nineteenth-century luminaries like Alexander von Humboldt, British polymath William Whewell, and Henry David Thoreau.[12] Thus, by bridging science and poetic discourse and their respective discourses and disciplines grounded in constructions of objective or subjective realities, Childs creates what Walls describes as "'relational knowledge,' not of subject against object but of the new whole that subject and object make together" (204). Such is the epistemological and discursive vision Childs embraces in *The Secret Knowledge of Water* as he spans disciplinary boundaries and recognizes the relationality of the knowledge that he acquires to construct a text motivated by various modes of expression to underscore paradox and the desert's materiality. Accordingly, Childs's experiences reflect what Walls discovers in Whewell's writings: the coming together of various perspectives "to form a new, coherent truth." Walls's consideration of this "truth" or enlarged vision, as I see it, is a useful approach to examine Childs's work and its representations of the Southwest's deserts. Adopting the model of a river to explain the convergence of divergent knowledge, Whewell explains that "the stream of knowledge from various classes of facts will constantly run together into a smaller and smaller number of channels; like the confluent rivulets of a great river, coming together from many sources, uniting their ramifications so as to form larger branches, these gain uniting in a single truth" (qtd. in Walls 205). Although Childs's work does not pursue a unified "truth," as did Whewell and other scientist-philosophers, his work nonetheless attempts to synthesize various epistemologies to create a more comprehensive and holistic understanding of the Southwest's deserts.

This model of a river's confluence provides an effective heuristic by which to understand Childs's perspectives on desert knowledge. From biologic data about *Triops* and cryptobiotic soil to the geographical surveys of a land commission and the stories gleaned from the Tohono O'odham and national park rangers, boaters, flash-flood survivors, and what the desert itself reveals, Childs creates a mosaic of "truth" about what the desert is and what it does. Unlike the unified reality Whewell imagines, Childs navigates his way through past and present knowledge of the deserts and their water, never positioning one approach as the definitive way to access this secret knowledge. And as I have suggested above, by drawing on these apparently paradoxical epistemological approaches, Childs allows the desert and the water that has shaped the land to have a voices of their own. By bringing all these voices into conversation within the text, Childs demonstrates that the desert does not just *mean* something but that it *does* something as well: It both exerts its own agency to shape human and nonhuman entities and reveals its knowledge that only those willing and patient enough—and perhaps even lucky enough—can begin to understand.

To conclude, I turn to Frank Stewart's observation about nature writers. He states, "Whether scientists or poets, nature writers make us aware that neither biology nor imagination by itself can illuminate the call of the last American timber wolf, [or] the tossing meadow grasses in a mountain rainstorm.... But both disciplines, working together, may give us a new, more powerful lens of perception" (xix). In the case of the deserts of the American Southwest, Childs's text provides a unique perspective by which to approach this region, an approach that brings Teorey's "story of health and prosperity" and Buell's "vision corrective" into play (7; *Writing* 246). By negotiating various discourses and their traditions, Childs resists trying to reduce our understanding of these deserts to one epistemological model. In doing so, he demonstrates a willingness to accept the gaps in his learning, thereby granting the desert a sense of autonomy and agency that transcend human comprehension. By deferring to the desert's and its water's own systems of knowledge, systems that continually prove enigmatic, Childs allows these entities to become an instructor of sorts, one that inspires, challenges, and refutes his previous assumptions and hypotheses about these unique entities. In a place such as the Colorado River Basin where human and nonhuman communities face an uncertain future, *The Secret Knowledge of Water* demonstrates that no one approach to understanding and addressing these current environmental issues is adequate. Through his implementation of paradox as a way to highlight desert water's unique properties, Childs speaks to multiple knowledges, which at times appear contradictory but demonstrate how to more sustainably reinhabit the watershed.

CHAPTER THREE

New Currents in Colorado River Boating Narratives

Westerns, Wilderness, and Women

Coming to the Canyon and immersing myself
in its beauty have brought a self-confidence and
appreciation I never knew I was capable of,
liberating me from the "baggage" of life.
—Denise Traver, in *Writing Down the River*, by KATHLEEN JO RYAN

LAVA FALLS IS PERHAPS the most storied rapid on the Colorado River and the pinnacle of any river runner's experience through the Grand Canyon. Formed by a cycle of prehistoric lava flows that have repeatedly dammed the river's progress and the Colorado's constant erosive power, this rapid has captured the imagination of countless river runners who have attempted to navigate this notorious stretch of water. For those river runners who look to capture their experience in writing, Lava provides the perfect climax for a river yarn. Situated near the end of most river trips through the Grand Canyon, Lava is a fitting high point, as the foreboding of and anticipation for this hallowed rapid culminate in a rush of pure adrenaline. Mirroring the river's progress downstream, a typical river narrative follows this linearity using Lava and one's passage through it as the culmination and resolution of a text.

Such is the case with Elizabeth Hyde's *In the Heart of the Canyon* (2009), a fictitious account of twelve vacationers who convene on the Grand Canyon for a two-week river trip with their trusty guides: JT, Abo, and Dixie. As in nearly all narratives about the Colorado River's stretch through the Grand Canyon, the rapids tend to take center stage. At the beginning of day 7, JT explains to the group, "It's not about the rapids," yet he knows full well that the beauty of the surrounding cliffs, the intimacy of the canyon's endless side canyons, and the surprising abundance of flora and fauna can hardly compete with the thrill of the white water to come. As Hyde writes, "Try convincing twelve people not to get too excited about running the biggest white water on the continent.... There's no getting around it: ninety-three miles downriver from Lee's Ferry, it's about the Big Ones" (170). Amid names like Granite, Hermit, and Crystal Rapids, Lava reigns supreme. Yet unlike so many tales that climax with the rise and fall of one's passage through Lava, *In the Heart of the Canyon* takes a unique departure from this formula. Anticipation peaks when the party awakes on day 11 to JT's announcement: "It's Judgment Day" (225). In preparing for Lava, everything goes as expected as the guides scout the rapid, the passengers give their life vests one last safety check, and they all hold on for dear life. And even when Amy, the obese, angst-filled teenager, falls overboard during her tumultuous passage through the rapid, the narrative seems on par with others. But when she's rescued and brought to dry ground in Lava's tail waters, the real excitement begins. The unexplainable health issues Amy has experienced throughout the voyage come to a head at the foot of Lava, as it is clear to all that the seventeen-year-old is in labor.

Shifting the climax of a Colorado River boating narrative from the excitement and adrenaline rush of pounding through rapids to giving birth, Hyde places her text within a growing body of boating literature that looks beyond this male-dominated genre that places adventure, wilderness, and individuality—elements of western myth and the western genre—at a premium. If, as Susan Lee Johnson argues, "no place has been so consistently identified with maleness—particularly white maleness—as the region imagined as the American West," then texts like Hyde's posit an alternative vision of the West for the reader to consider (495).[1] In fact, as recent recovery projects of women's writing in and about the West attest, women have been integral to the rise of the genre, despite their marginalization. Victoria Lamont's *Westerns: A Woman's History* (2016), for example, notes how Emma Ghent Curtis's 1889 cowboy novel, *The Administratrix*, predated Owen Wister's classic *The Virginian* (1902) by more than a decade. As Lamont explains, "Women were active at every turn during the period, between roughly 1880 and 1940, when the American frontier myth, after years of ghettoization in the dime novel, was supposedly 'reborn' as a dominant myth of American identity" (1). She then examines how it was not until the 1920s that the genre became the sole territory of male writers (125).[2]

This chapter continues this reclamation of women's voices in the West by focusing specifically on narratives about boating the Colorado through the Grand Canyon, perhaps the most iconic and celebrated stretch of river in the United States. As the women grapple with the traditional topoi—the recurring literary conventions and themes—that have come to define Grand Canyon boating literature for roughly 150 years, they posit a vision of the New West that embraces a more complex understanding of this mythologized region. For women's voices to reorient our understanding of the West, they must, as Krista Comer contends in her critique of the Old West as "wild landscape," "first expose it if they are to avoid being its victims. But to then go further and to survive with any kind of female subjectivity intact, they must reemplot it" (160).[3] This manipulation of the genre by positing alternative topoi is also echoed by Kristine McAndrews in *Wrangling Women* (2006), her study of the horsewoman in rural eastern Washington. Examining how the stories these women tell negotiate their own marginalized position in a traditionally male line of work, McAndrews explains, "When telling their stories, women in this community employ traditional male narrative techniques but stretch or undermine these strategies by introducing nontraditional images and themes" (xiii). Such is the case with the interviews, personal narratives, essays, novels, short stories, and poetry that represent the range of genre choices in women's Grand Canyon river writing that has proliferated over the past few decades.

As a result of changing gender expectations, women have had greater access to outdoor recreation as evident in Hyde's tale, and the proliferation of river narratives along the Colorado and its major tributaries starting in the 1990s reflects this shift. These include Ann Weiler Walka's *Waterlines* (1993), Ellen Meloy's *Raven's Exile* (1994), Louise Teal's *Breaking into the Current* (1994), Patricia McCairen's *Canyon Solitude* (1996), Kathleen Jo Ryan's *Writing Down the River* (1998), Katie Lee's *All My Rivers Are Gone* (1998), Laurie Wagner Buyer's *Side Canyons* (2004), and Heather Hansman's *Downriver* (2019).[4] These examples represent women's voyages down the Colorado River and its tributaries in ways that employ techniques outlined in Jane Tompkins's *West of Everything* (1992) and Kristine Groover's *The Wilderness Within* (1999) to reemplot the Colorado River western: the emphasis on action as a reaction against the constraints of domesticity, the aversion to talk, and the "monolithic austerity" of the landscape (Tompkins 77), as well as what Groover identifies as the pursuit of the "spiritual quest," defined by "heroic protagonists [who] undertake physical journeys whose destination is a greater understanding of or connection to the spiritual world" (1). While this quest "is a tradition which pointedly excludes women," Groover clarifies that women also enact their own form of the spiritual quest evident in women's experiences in the realms of "domesticity, community, and storytelling" (3, 11). Although many of the women who travel to the river perform a very similar quest to that evident in men's Colorado boating narratives as they seek out some of the nation's most rugged and iconic landscapes for personal enlightenment and to comment on current environmental concerns along the river, they nonetheless do so by complicating both the typical male approaches and the essentialized triad of female discourses Groover prescribes for women writers entering the wilderness. Their accounts of this place, in Comer's words, "evidence female attempts to access public space, public discourse, public issues, public life, and public power" (28). As a result, they offer us alternative means by which to understand how the river shapes and is shaped by those who encounter it.

To suggest how women's Grand Canyon boating narratives offer us more complex understandings of the canyon and broader watershed, I turn to selections from Teal's *Breaking into the Current*, McCairen's *Canyon Solitude*, Buyer's *Side Canyons*, and Ryan's *Writing the River* to embark on an initial foray into women's Colorado River westerns and, specifically, those focused on the river's stretch through the Grand Canyon. In doing so, I wish to underscore how a woman's voice is not an anomaly within this genre, but part of a figurative "gathering of waters" where many women are adding to our understanding of the river and canyon. Of course, such a broad treatment of texts runs the risk of diminishing their contributions. However, my intent here is not to be exhaustive but rather illustrative. Through an examination of these texts' various discursive approaches that span poetry and interviews, essays and memoir, and their engagement with and negotiation of the canonical Colorado River western's topoi, I argue that these women complicate the public-private representations of the river and canyon, as well as the cultural constraints they face because of their identity. By appropriating the western genre and its associated topoi, which include "the encounter with wilderness, the excitement of danger and challenge, the violent act of confrontation and commitment, the final slow surrender of freedom to advancing civilization," these women enter a well-recognized literary body through which to recount the adventure, learning, and wonder that they experience on the river (Armitage 16). Yet it is exactly via this appropriation and manipulation of the genre's poetic discourse that these narratives become clearly rhetorically motivated as they seek to justify the women's presence on the river to carve out a place for their voices within this traditionally male-dominated activity and literary tradition. Negotiating such themes as domesticity, community, storytelling, and empowerment, these texts posit a vision of the New West that values the contributions of tributary voices. Embodying this position, their work becomes incredibly instructive as it questions Old West ideologies that have made the river an exclusionary space and gives rise to the kind of awareness and open discourse needed to more equitably address the needs of this threatened resource.

Colorado River Westerns

Prior to examining how these women boaters negotiate canonical Colorado River topoi, I first turn to the Colorado's boating history and the earliest narratives inspired by downriver journeys to establish the genre's conventions. In 1867 John Wesley Powell spent a summer in the Longs Peak region of Colorado Territory, exploring the headwaters of the Grand River (later the Colorado) and speculating about its journey through the canyons of the West. These early imaginings eventually gave way to his famed expedition two years later, which aimed at "penetrating still farther in that canyon country" of the Grand, Green, and Colorado Rivers (*The Exploration* 117). On May 24, 1869, Powell and company pushed the *Emma Dean* and three other boats from the shore of what would be later named "Expedition Island" and turned his back on the small outpost of Green River, Wyoming. Over the course of the next three months, Powell and his crew drifted south through the canyons of the Green and Colorado Rivers, charting new territory (to them) and opening up the marvels of this region to the masses.

For Powell and his men, this expedition was a journey into the unknown, an opportunity to put their names alongside the great explorers of the continent who had gone before them, returning triumphant in their descriptions of sublime landscapes, perilous rivers, and little-known cultures. Yet the realities of their journey down the Colorado challenged any visions of grandeur they may have imagined. While Powell's journal chronicles the marvelous places they encounter as they race downriver, his account also reflects their uncertainty, reservation, and even despair as his crew faces countless hardships navigating a dangerous, unpredictable river. Perhaps Powell's most famous entry captures the scene as the expedition enters what would later be named the Inner Gorge of the Grand Canyon. Having already spent months fighting perilous rapids and intense heat, the major confides, "We are now ready to start on our way down the Great Unknown.... We are three quarters of a mile in the depths of the earth, and the great river shrinks into insignificance as it dashes its entry waves against the walls and cliffs that rise to the world above.... We have an unknown distance yet to run, and unknown river to explore. What falls there are, we know not; what rocks beset the channel, we know not; what walls rise over the river, we know not" (247). Whereas the group had traveled through previously explored country, their journey into the Grand Canyon was a step into the void—a blank spot on the nation's maps. Concerned about food rations, the distance they still had to travel to reach the Mormon settlements along the Virgin River, and the unknown dangers ahead, Powell's outlook is far from the confident, optimistic mood that initiated this fateful voyage.

Although Powell's initial account garnered a lukewarm reception in 1875, subsequent revisions such as the 1895 *The Exploration of the Colorado River and Its Canyons* made this a classic of western American literature, which has played a paramount role in the way that the river has since been imagined. Donald Worster explains that Powell's record "became the most popular literary product of all the western surveys. Over the next century no one among his contemporary explorers and surveyors...was as widely read as Powell" (*River Running West* 332). One of the reasons for the staying power of the 1895 account is its emergence during a unique historical moment when the American frontier captured the nation's attention and western romance and the dime novel ruled the literary marketplace. Two years prior to *The Exploration*'s publication, Frederick Jackson Turner addressed the American Historical Association at the Columbian Exposition in Chicago and boldly declared that the frontier had officially closed. Ironically, while one frontier supposedly closed, another one steeped in a mythic construction of this bygone era emerged. While nostalgia typified America's understanding of and engagement with the West long before Turner's proclamation, it thrived as a pervasive cultural force within the western genre and the dime novel during the latter half of the nineteenth and early decades of the twentieth centuries. Relying on the fundamental reproducibility of scene, character, and action, Powell's boating narrative embodies a particular set of narrative structures and topoi that subsequent Colorado River accounts would reproduce again and again (Brown 6). As a result, Powell's record stands at the fore of a highly conventionalized albeit minor genre of western literature that continues to shape public engagement with the river more than 150 years after his first voyage.

Recognizing the potential market for river narratives that capitalized on the widespread interest in the American West, others who had accompanied Powell down the Colorado provided their own side of the story. In the same year that Owen Wister published *The Virginian* (1902), the quintessential western, Frederick Dellenbaugh, a member of Powell's second expedition in 1871–72, published his *Romance of the Colorado River*, which chronicles the river's "discovery" in 1540 by Spanish explorers through his own journey with Major Powell. Emphasizing the distinction of Powell's two voyages as the first to travel through "the whole line of canyons," Dellenbaugh writes, "If danger, difficulty, and disaster mean romance, then assuredly the Colorado of the West is entitled to first rank, for seldom has any human being touched its borderland even, without some bitter or fatal experience" (vii). By charting the exciting history of encounters with the river over hundreds of years by conquistadores and mountain men, and by referring to Powell as "the One-armed Knight" who made "A Bold Attack on the Canyon" (184), Dellenbaugh extends Powell's narrative framework and themes of challenge, struggle, and eventual survival into a far more melodramatic realm depicted by a long trajectory of male heroism and conquest of this mythic landscape. His penchant for sensationalism is clearly evident from the opening pages in which he depicts the river as "a veritable dragon, loud in its dangerous lair, defiant, fierce, opposing utility everywhere, refusing absolutely to be bridled by Commerce, perpetuating a wilderness, prohibiting mankind's encroachments, and in its immediate tide presenting a formidable host of snarling waters whose angry roar, reverberating wildly league after league between giant rock-walls carved through the bowels of the earth, heralds the impossibility of human conquest and smothers hope" (2). Dellenbaugh's retelling of the voyage takes advantage of the common literary topoi of adventure, peril, and conquest to promote his fantastic tale. Likewise, by referencing those significant places noted by Powell such as Echo Park, the Dirty

Devil, Labyrinth Canyon, Music Temple, the Crossing of the Fathers, and Sockdolager Rapid, we begin to see the discursive construction of the river become rather stylized and conventional through the repetition of these topoi and the poetic, rather than scientific, discourse that underlies them. This particular formula's coalescence also depends on portraying the river as an exclusive physical and imaginative locale: The river is a place only for those men courageous and skilled enough to face these perilous waters and then write and publish their experiences.

Captivated by Powell's and Dellenbaugh's accounts and armed with knowledge of three other voyages down the Colorado, two brothers, Ellsworth and Emory Kolb, arrived in Green River City, Wyoming, in early September 1911 to record the first moving pictures of the Green and Colorado Rivers. Prepared for what they called the "Big Trip"—a journey that carried them from Green River City to Needles, California—they hoped to "bring out a record of the Colorado as it is, a live thing, armed as it were with teeth, ready to crush and devour" (4). Echoing Dellenbaugh's portrayal of the river as a dragon, the Kolbs' own voyage became a direct outgrowth of the nascent genre of Colorado River boating literature. When the Kolbs landed in Needles on January 18, 1912, they had achieved their dream. The following year, Ellsworth spent eight days floating from Needles to the Gulf of California. Thus, their combined expedition from Wyoming to Mexico over the course of 101 days culminated in the first recorded journey from Green River City to the Gulf while also providing the first film images of this remote corner of the American West. Desiring to publish an account of their own in order to add their voices to those of the revered explorers who preceded them downriver, Ellsworth produced *Through the Grand Canyon from Wyoming to Mexico* (1914).

Building on the genre established by Powell and further dramatized by Dellenbaugh, Kolb's narrative relies on many of the structural and thematic elements of the western to promote their experience. Kolb's text also had the good fortune of securing an endorsement from Wister himself to further entrench the Colorado boating narrative within the western genre. Providing the text's foreword, Wister evokes quintessential western themes as he introduces the Kolbs' journey through the canyons of the Southwest: "Every youth who has in him a spark of adventure will kindle with desire to battle his way also from Green River to the foot of Bright Angel Trail; while every man whose bones have been stiffened and his breath made short by the years, will remember wistfully such wild tastes of risk and conquest that he, too, rejoiced in when he was young.... This place exerts a magnetic spell.... Bend after bend this trance of beauty and awe goes on, terrible as the Day of Judgment, sublime as the Psalms of David" (ix, xi). Wister's lofty and melodramatic description of the Grand Canyon, noted by its references to its mystic qualities, unsurpassed beauty, and the Kolbs' journey through this region, speaks to the era's praise of explorers and heroic deeds that pit man and nature against one another. His words also firmly embed the river-running narrative as a subgenre of the western, making subsequent river narratives heavily steeped in mythic landscapes, nostalgia, and perilous adventure for the male subject—a motif that extends through a long line of river narratives by men that include Harold Leich's *Alone on the Colorado* (1938), Edward Abbey's "Down the River" chapter from *Desert Solitaire* (1968), John McPhee's *Encounters with the Archdruid* (1971), Colin Fletcher's *River* (1997), Jonathan Waterman's *Running Dry* (2012), and Kevin Fedarko's *The Emerald Mile* (2013).[5]

Among these, Abbey's essay stands out for its deliberate sexualization of the river. His descriptions recall Annette Kolodny's observations from *Lay of the Land* in which she speaks to "America's oldest and most cherished fantasy: a daily reality of harmony between man and nature based on an experience of the land as essentially feminine." Noting this pervasive view of the American landscape, Kolodny clarifies that this feminization of the land has led to its exploitation. She writes that we have looked at the land not only "as mother, but the land as woman, the total female principle of gratification—enclosing the individual in an environment of receptivity, repose, and painless and integral satisfaction" (4). This two-sided perspective of the land as both lover and nurturer typifies Abbey's experience as he embarks on his voyage downriver "to see [Glen Canyon] as Powell and his party had seen it, not knowing what to expect, making anew discoveries of others" (*Desert Solitaire* 195). Stepping away from shore, Abbey paddles his way into the main current and "onto the brown silt-rich bosom of the Colorado." Effortlessly gliding downriver, Abbey's earlier hesitation turns into bliss: "My anxieties have vanished and I feel instead a sense of cradlelike security, of achievement and joy, a pleasure almost equivalent to that first entrance—from the outside—into the neck of the womb."

As if Abbey's reverie isn't enough, he goes on to more explicitly note that they were "indeed enjoying a very intimate relation with the river." He continues, "I am fulfilling at last a dream of childhood and one as powerful as the erotic dreams of adolescence—*floating down the river*" (191). He concludes this sexual fantasy appealing to a male audience he believes would perfectly understand this experience. "Every man that has ever put forth on flowing water," Abbey surmises, "knows what I mean" (192). While Kolodny does not mention Abbey's text in her study, his work nonetheless stands as an exemplary case of the fantasy and desire that define much of America's interaction with nature. This conflation between women and nature "assume[s]," as Connie Bullis states, "that the oppressions of women, races, classes, and nonhuman nature are interconnected parts of the same dynamic" (124). Thus, as Abbey describes the Colorado as an exotic female whose dark flesh awaits him and who eagerly succumbs to his sexual desires, there is no consideration of how this female presence responds to such entreaties. The entire description emanates from Abbey's perspective, which suggests that his "lover," while highly prized, is not an agent unto herself. With no voice, the river is only something to possess to satisfy the explorers' desires.

The appearance of these foundational Colorado River narratives and their poetic discourse and common western topoi are by no means coincidence. Recognizing the nation's fascination with the American West captured in this popular medium, these authors rely heavily on the formulaic nature of the western to celebrate this wild region, construct their own heroic tales, and garner widespread popularity. From Powell's earliest account through the anticipatory nostalgia in Abbey's writing as he mourns a doomed Glen Canyon, Colorado River narratives have adopted the western's and dime novel's formulaic pattern and replication of plot, scene, and character to create its own unique genre. These river westerns position men at the center of the text, chart its exploits navigating unpredictable and often unknown waters and territories, describe the allure of remote sublime lands on the protagonists' psyche, and marginalize community and communicative experiences in favor of isolation and quiet.[6] These characteristics, coupled with specific references to key geologic and cultural sites along the river, demonstrate the writers' reliance on poetic discourse to convey engaging tales of their river experiences. However, like the dime western that is principally concerned with "men writing about men" (Brown 32), the Colorado River boating narrative is equally concerned (whether consciously or not) with delineating the river-running experience as a male pursuit. Thus, the pervasive poetic discourse that harnesses these common topoi gives way to the rhetorical, which delineates boundaries and sections of this genre as the offspring of an exclusive discourse community. As such, these narratives both ignore and repress a female presence on the river and within the literary tradition of the Colorado River boating narrative.

However, a closer look at the river's literary history reveals how women, beginning in the mid-twentieth century, adopted, nuanced, and repurposed the Colorado River western to create a new literary version of the river and boating culture. Edith Kolb traveled a portion of the river with her brothers and is believed to be the first woman to run a major rapid (Lavender, *River Runners* 62). Bessie Hyde ventured down the Grand Canyon with her husband, Glen, on an ill-fated honeymoon trip in 1928 when both drowned. Mary Remsen North, a ten-year-old, traveled from the Boulder Canyon (Hoover) dam site to Yuma, Arizona, and through the delta with her parents, publishing her account in 1930.[7] Elzada Clover and Lois Jotter traveled through the Grand Canyon in 1937 as part of a University of Michigan–sponsored botanical survey to become the first women to navigate this stretch of the river (96–97). A year later, Genevieve DeColmont joined her husband, Bernard, and friend Antoine DeSeyne to kayak down the Colorado, becoming the first woman to pilot her own boat on the river, much to the fanfare of her native France (Webb 104–5). Other notables include Doris Nevills, Mildred Baker, Katie Lee, Ann Zwinger, and Georgie Clark.[8] Lee, Zwinger, and Clark, in particular, left a significant impact on the river through their literary production and for their pioneering efforts on the water. Lee's *All My Rivers Are Gone* (1998) chronicles her excursions in Glen Canyon beginning in the 1950s and her vehement opposition to Glen Canyon Dam. She remained, up until her death in 2007, one of the most ardent advocates for a free-flowing Colorado. Ann Zwinger, the well-known nature writer, penned *Run, River, Run* (1975) and later *Downcanyon* (1995) to establish herself as one of the keenest observers of the American Southwest's flora, fauna, and topography. Clark's longstanding presence on the river, beginning in the 1950s, broke new ground in guiding river trips that continues to shape the industry today. Clark captures these pioneering efforts in her autobiography, *Georgie Clark: Thirty Years of River Running* (1977).[9] It is within this tradition that the following texts, beginning in the mid-1990s, take their own place as significant contributors to the history of the river.

Teal's *Breaking into the Current*

Louise Teal's *Breaking into the Current* is neither a novel nor a memoir, like so many other Grand Canyon boating narratives, but a collection of interviews with twelve female Grand Canyon river guides who address community, domesticity, and empowerment through the stories they tell. Such a departure from the genre's usual poetic (that is, narrative) discourse is typical of much women's writing, as numerous scholars have argued. Susanne Bounds and Patti Capel Swartz explain in their consideration of women's desert writing, "The Western desert woman writer is developing a genre of her own, one often not solely definable in terms of genre categorizations of poetry, prose, fiction, nonfiction or drama" (77). In *The Desert Is No Lady* (1987), Vera Norwood and Janice Monk extend this observation of how women writers challenge traditional genre conventions, suggesting that "to understand women's responses often we must seek out lesser-known works of literature and art, folk forms such as weaving, pottery, embroidery, and quilting, and women's diaries, journals, reminiscences, and oral histories" (4). Even though these narratives draw on the Colorado River western's topoi as in other works, *Breaking into the Current* is a more rhetorically motivated text than those considered later in this chapter, as Teal's collection more directly tackles the issue of sexism on the Colorado and the means by which her interviewees address this discrimination while guiding.

Teal, a river guide and writer, captures the voices of women who have made their way into a traditionally exclusive club of male river runners in the Grand Canyon. Dedicating the book "to all women navigating the changing currents of our time," she acknowledges Georgie Clark's influence in opening the canyon's recreational opportunities to other women and then turns to the experiences of those women who followed Clark in subsequent decades through 1990, when only 16 percent of the canyon's numerous commercial rafting companies employed women as full-time guides. Despite the advances that women made in many other industries during this period, guiding in the Grand Canyon was still very much a man's domain. However, when compared with the 1970s when Teal first began guiding, this meager percentage actually reflects substantial gains for women within the guiding business. Then, she says, "You could count the number of women guides...on one hand" (xii). Within this context Teal works to capture the unique contribution that women have made within the canyon and the boating community.

Challenging the perceptions about the Grand Canyon, which is, in Teal's words, "the last bastion of the male river god," she emphasizes stories and experiences by female guides that speak to the unique position women have on the river (51). These passages both align with and challenge western themes that aptly define this traditionally male-dominated profession. For Teal, gathering stories is an integral part of the river experience that demands attention and a willingness to see beyond traditional speech acts to discern meaning. She explains, "When you stick your oars in the water, you're feeling the whole story. There's no words, but it's the language of the formation of the earth" (59). Teal's engagement with the river speaks to a relationship where more than language is at play in causing the Grand Canyon's story to surface. If people are observant and willing to listen to the lessons the canyon and river have to share, they will come to understand the canyon's geologic wonders. Similarly, as one dips a paddle into this proverbial river of words coursing through the Grand Canyon boating genre, the stories of those who have passed before emerge. Through her text, Teal seeks to bring to the surface the "whole story" of these pioneering boatwomen whose river experiences have largely gone unnoticed but that add necessary depth to understanding how the river has shaped those who make their living by reading its features.

River employment is one of the significant narratives that Teal shares in the text to delineate how women's river work differs from that typified by men as embodied in the western. Jane Tompkins identifies work and action in the western as another theme that reasserts the Old West. While she suggests that westerns have functioned as an escape from everyday life to which work would belong, she argues that "hard work is transformed here from the necessity one wants to escape into the most desirable of human endeavors: action that totally saturates the present moment, totally absorbs body and mind, and directs one's life to the service of an unquestioned goal" (12). Such work takes on this crucial aspect in the western because it is this struggle against one's surroundings, whether human or nonhuman, that gives purpose to one's life. As Tompkins elaborates, "The trouble with ordinary work isn't, as people generally assume, that it demands too much of you but that it doesn't demand enough" (15). The western, on the other hand, provides action that requires one's utmost attention and focus. Life in the western depends on the ability to toil and sacrifice, to "satisfy... a hunger not for adventure but for meaning" (15). Although most river narratives by men reference the work involved in piloting watercraft downriver, much of this is cast within the framework of recreation—the toil for pure pleasure rather than employment to support one's day-to-day living.

Teal emphasizes the role of work in the lives of her female colleagues to show how women are as capable of doing this male-dominated job and similarly interested in finding greater purpose for their lives. "Working long hours," the female guide "doesn't punch out at five o'clock each day during the 225-mile trips that last eight to eighteen days," Teal observes. In fact, she informs her audience, "Sometimes we row against seemingly endless upstream winds. Our skin is dried out by too much sun, and our brains are fried by hundred-degree heat. We face other assaults on our bodies, the largest being some incurable form of foot rot" (1). But for the women who routinely endure these particular hardships, this is a small price to pay for being in the canyon. Like the western hero who finds true purpose for living through his unique labor, these women find meaning to their lives and come away with a greater sense of self as they take on challenges traditionally faced by men.

Teal's description of work also challenges the exclusivity of the guiding profession in a way that establishes a collective identity among the disproportionately few women in this industry. Marilyn Sayre, a predecessor to Teal in the canyon, recalls her experiences with some of her male coworkers: "I felt, although they never said anything overtly, that some of them did not want me there, that they would rather have another guy. It was the first time in my whole life that I ever really experienced what I'd call male chauvinism.... Maybe I was a threat to them, and they couldn't warm up to me. It made it hard for me to work down there" (31). This reaction is not an isolated case. Buzz Holmstrom, one of the pioneering boaters in the Grand Canyon, once remarked, "Women have their place in the world, but they do not belong in the Canyon of the Colorado" (Lavender, *River Running* 94). Such opinions about women's presumed inferiority exacted both emotional and physical tolls on the female guides. Yet these conditions helped solidify the spirit of sisterhood that Teal and others invoke throughout the text, which leads her to claim that "these women speak for us all" (Teal xiii). Thus, this text is far from just a collection of adventure narratives, the usual fare so readily available in a quick perusal of the Grand Canyon's boating literature. Instead, it represents a powerful rhetorical act, one that strives to carve out a special place for women's experiences both on and off the river, further evidenced by Teal's admission that this book is not only for those intrepid women of the river who "only follow[ed] their hearts amidst the pressures and possibilities facing all women in our particular era," but an homage "that honors all women" (xiv).

At the same time, her description of the working conditions also functions to establish her credibility as an insider—to show that she and the other women in the text know what it takes to be successful as a river guide. Only those who have awakened at four thirty in the morning to brew coffee, cooked over a stove in 110-degree heat, continually loaded and unloaded multiple rafts, attended to the group's toilet needs, led hiking expeditions, and dealt with tourists' wide range of emotions and questions, all while being the ever-gracious host, know the unique challenges of this line of work. However, Teal also recognizes that complaining does little to improve such difficult tasks and one's status among the other guides. Since, as her fellow guide Liz Hymans acknowledges, "boatmen seemed to thrive on an atmosphere of competition, . . . so you had to look like you were made of steel the whole time," Teal is quick to move beyond the challenges of a guide's work to provide the reader with a more favorable depiction of this labor that captures the real reason she, and others like her, continues to return to the river year after year (43). "But for the six-month river season," she writes, "these minor irritations are more than offset by watching that massive pile of rock, the Grand Canyon, work its particular magic on folks" (1). From the accounts that follow, the canyon and the river's "magic" work not only on those paying customers but also on the subjects of Teal's book, who, with her, "were all in the same romance" (xiii). The work performed in leading commercial trips, as well as the river itself, is not a chore but a delight for these women, a unique occupation whose benefits are paid in uncompromising beauty and lifelong relationships rather than high wages and stock options.

By entering into the discourse of work cast within an idealism of hard labor, Teal engages a key western topos while also emphasizing the unique perspective of a female river guide. This confluence with and divergence from the western formula continues throughout the text as Teal emphasizes other stories that speak to the unique experiences of women rafting the Colorado through the Grand Canyon. Drawing upon the western's celebration of the heroic figure, Teal includes stories that position the female guide as hero in the eyes of other guides and passengers as she goes about her duties to ensure that all under her care have safe passage through the canyon. Such a move places these women on equal level with men to debunk the "myth…that lingers still, that it takes a large, powerful man to row a boat through whitewater," as Patricia McCairen explains in her own narrative considered in the next section (152). A representative example comes from Teal's retelling of Suzanne Jordan's memorable run at 24.5-Mile Rapid in the historic high water of 1983. After flipping her boat and being tossed through another rapid a short way downstream, "it looked like she had jumped on someone's upright boat, tied the two flipped rafts to it, and was rowing all three boats to shore." Suzanne's fellow guide, David, observed the entire event and marveled at the strength of this petite woman to recover the rafts. As Teal writes of David's reaction, "I had a hard time rowing one boat to shore at 62,000 [cubic feet per second], and here she was pulling two upside-down boats to shore" (85).[10] Such a statement demonstrates not only that women are equal to the challenges of river work, but that they can exceed men's abilities.

But strength alone does not fully describe the heroic qualities that Teal emphasizes in her subjects. When a flash flood in Havasu Canyon overwhelms a passenger and a guide who goes to rescue her, it is Suzanne who is in the right place at the right time with the rope to throw out to the two people being swept out of the canyon and into the Colorado. David, the guide she rescued previously, praises Suzanne, stating, "She's always spot-on and one of the bravest people I've ever worked with down here" (89). This courage and foresight demonstrated in this example are indicative of the rest of the women throughout Teal's text. As every western has its hero, so do these Colorado River westerns as they emphasize the heroic feats accomplished along the river. Yet Teal's purpose is to create new heroes through the stories she tells. Speaking of her fellow female guides, one boatwoman remarks, "They were neat to aspire to be like, that maybe you didn't have to be macho and look like a guy and act like a guy to be competent" (170). Very much a text about women for women, *Breaking into the Current* provides a valuable alternative narrative to the typical male version even as it reproduces some of its principal features. While the women in this text tap into similar topoi of the western-like adventure, a love of grand landscapes, and hard work, they do so from their own experiences, emphasizing how they have had to struggle against pervasive stereotypes and prejudices to create their own space on the river. As she concludes her text, Teal explains, "It is wonderful to watch another generation of women—young women who have grown up hearing river stories in which the storytellers are women and the heroes are heroines" (172). The stories this text tells promote a new version of the western hero, one that is every bit as courageous and capable as her male sidekicks.

McCairen's *Canyon Solitude*

Two years after Teal's publication, Patricia McCairen's *Canyon Solitude* joined this growing body of female boating narratives to recast the western hero, as she appropriates and refashions the Colorado River western to explore her own insecurities and joys as a single middle-aged woman and to ultimately argue for the legitimacy of a woman's experience on the river. This memoir chronicles McCairen's journey rafting the Grand Canyon solo and the experiences that led up to this momentous event in her life. She begins her tale *in media res,* perched on a ledge in the heart of the Grand Canyon, unable to move for fear of falling. Revealing her fear of heights, she draws a connection to the same inaction in her life that has stalled her progress toward becoming what she feels others want her to become. She confides: "The same fear that has kept me stuck in life so many times before, afraid to move forward, to take a step that would free me from the ordinary, the mundane, the insufferable. A crippling fear that deadens my potential and limits my relationship with the world. It's so easy to cling to the familiar, even when it's deplorable." Yet in the midst of such paralysis, she finds the will to move and slowly inches her way to safety, knowing that "I've taken more difficult steps in life, chosen pathways that required more of me than this single step" (5). She concludes the opening chapter, "The canyon has spared me once again" (6).

These three passages establish the direction of McCairen's text and reveal her motivation for being on the river. Within them we find the traditional western boating topoi of being challenged by nature, facing fear, and achieving ultimate triumph. At the same time, however, McCairen's sentiments embody a sense of humility not often seen among western protagonists. Unlike the undaunted hero prepared to face any challenge, she recognizes her fears, which have long plagued her both in and out of the canyon. McCairen's fear is a deeply felt sense of failure and inadequacy resulting from her inability to live up to others' expectations and her own dreams. It is this kind of fear that leads her to remain in the same rut day in and day out, choosing safety and sadness rather than risk and possible freedom. Far from the uncertainly and trepidation that Powell and his crew face as they venture through a rapid, McCairen's fear reveals a vulnerability unlike anything the male narrators suggest. Out of these opening pages surfaces a work that is at once ready to invoke a traditional male genre and reassert its topoi while also using those same elements to comment on the woman's place in society and on the river. This continual negotiation between perpetuating a well-established formula and repeatedly critiquing it establishes McCairen's text as one of the foundational Colorado River boating narratives that places a woman at the center of the text rather than a man. McCairen is ever cognizant of her marginalized position as a woman and uses her trip down the Colorado with its trials and triumphs as a larger metaphor for how to challenge the status quo to create a new vision for women's opportunities on and off the river.

To urge this change, *Canyon Solitude* perpetuates a number of male river western conventions. Along with the numerous discussions about adventure and sublime topography, McCairen feminizes the river in a complicated way that aligns with both Abbey's sexualized portrayal and Kolodny's earth-as-mother topos. McCairen describes the Colorado as "a sleek, beautiful goddess, alluringly seductive, forgiving to those who love her," a river that "entices" and is "beguiling" (51, 52). However, earlier passages reveal that the river is less a lover than a nurturer. She refers to the Colorado as "Mother River" and endows it with strength rather than the passivity that defines Abbey's river (22). McCairen explains, "She is a power that teaches those who open themselves to her. Her lessons may be subtle, or frightening. She cares not if we learn: It is up to us to seek her out" (51). While her representation of the river reproduces what has traditionally been a male-dominated cliché of Mother Earth, McCairen's use endows the river with greater autonomy to build on Teal's view of the river as teacher. Thus, a continual struggle emerges throughout this narrative between McCairen's need to enter a male-dominated genre to tell her story and her appropriation of the topoi to make her unique voice heard.

The reliance that McCairen has on this genre and its conventions continues as she recounts her voyage down the Colorado. Traveling downriver, McCairen finds that her ability to see the canyon and reflect on her experience is highly mediated by the men preceding her downriver. As she comes to Red Wall Cavern, she references Powell's and Abbey's perspectives on this famous alcove and notes an arch near 36-Mile Rapid named by the Kolbs as "the Bridge of Sighs" (87, 104). Likewise, when she reaches the ancient granaries in Nankoweap Canyon, another anticipated stop for Grand Canyon river trips, she references Fletcher's account of the area captured in one of his books. Attempting to re-create the scene of those who lived in the canyon hundreds of years earlier, she admits, "All I see is Colin Fletcher doing it" (115). These references to those sites made famous by previous river runners speak to McCairen's ongoing negotiation of her own experience on the river. Struggling to find her own voice and tell her own story, she often finds her journey mediated by men whose writings continually remind the author that she is an interloper in this male domain.

Notwithstanding the abundant physical and imaginative reminders that men have shaped how to see and experience the river, McCairen harnesses the genre's poetic discourse to posit alternative ways of knowing the river to create a complex, rhetorically motivated narrative. Her writing demonstrates a continual give-and-take between the male genre and her manipulation of it as she argues for a woman's rightful place on the river and desire to pursue her dreams even if contrary to societal norms and expectations that have traditionally assigned a woman's place within a domestic, communal sphere. A powerful example of this appropriation and critique emerges in her discussion of her first visit to the Grand Canyon and river, which initially locates her experience squarely within the predictable retreat-to-wilderness-for-rejuvenation theme. Toward the beginning of the narrative, she states, "I was born on the Colorado River at the bottom of the Grand Canyon, thirty-five years after my initial birth in New York City" (12). An avid traveler to various parts of the world, McCairen decides to follow the suggestion of a friend to spend her next trip on a guided raft tour down the Colorado through the Grand Canyon. With few expectations, McCairen soon finds her descent into the canyon enchanting, leaving her "totally overwhelmed the farther I descended" (16). Once on the river, the rafting party soon encounters its first of many rapids that mark the first day's journey downriver. Plummeting through Horn Rapid, McCairen explains, "Something in me snapped. With each motion I screamed—not a cry of fear or pain but of utter joy, released enthusiastically, naturally. I was vulnerable and open and totally happy. I was a child again, uninhibited, wild and free, riding a roller coaster with the excitement and anticipation of hanging on the brink before plunging down the near vertical slope" (20). McCairen's reaction to her experience descending into the canyon and crashing through rapids is predictable in terms of the poetic discourse she uses to praise the area's wilderness condition that presently defines the nation's fascination with remote, wild places. And to some degree, the outcome of this experience is similarly expected; she, like so many others who have flocked to the West in search of fulfillment, decides to abandon her "civilized" life in New York for "the rivers of the West and a tipi on Colorado's high plateau" (23).

Despite the apparent conventionality of this flight to the West in order to find herself anew, McCairen's rebirth signals a complicated engagement with the western genre's treatment of identity and the reassertion of masculinity that derives from journeying downriver. For Powell, Dellenbaugh, Kolb, and Abbey, a river adventure implied heading into the unknown, testing oneself against nature, and retreating from civilization's ailments. For McCairen, the rationale is somewhat similar. Yet unlike these men who often journeyed for the sake of fame and adventure, she seeks the river to deal with an inner crisis, one that is perhaps best understood in light of Colin Fletcher's justification for heading down the river on his own solo trip a few years previous to McCairen's. He explains in *River*, his own account of the journey, that he went because "I...realized that I'd grown soft. Things had been going too well lately. Too easily. I needed something to pare the fat off my soul...to make me grateful, again, for being alive" (7). Echoing Abbey's critique of civilization, Fletcher looks to wilderness to restore his vitality. McCairen similarly looks to the river to facilitate this transformation, but, again, the rationale is different. Rather than enjoying a cushy life where a trip to the wild can possibly remind Fletcher of his manliness, McCairen's desire for the wild is far more than a needed vacation from the daily rat race. Granted, she enjoys a job that allows her to spend weekends in Europe and vacation throughout the world. But what she realizes on her first trip through the canyon is that she's been living a life not wholly her own.

As a result of experiencing the "wildness and freedom" of the canyon and river that "continued to haunt" the author, McCairen begins to take control of her life, something that she had not done previously as she lived according to others' expectations of her (22). Only when she finds herself crashing through Horn Rapid does this new sense of being emerge. Naming this newly born version of herself Babe, McCairen reflects on this raft trip: "Something deeper had happened, something I didn't understand. A stranger had emerged in the canyon, and she fought to remain present. I tried to stuff her back wherever she had come from but she wouldn't have it. Babe had been born on the river and she demanded attention, she insisted on recognition.... Babe didn't want to be a sexy girl with bleached blond hair, makeup and short skirts to show off legs that turned men's heads.... She didn't care about being acceptable to others or pleasing men at her own expense" (22–23). Reminiscent of the woman trying to escape the confines of the yellow wallpaper in Charlotte Perkins Gilman's eponymous short story, McCairen's Babe emerges from the rapid's froth to challenge the traditional topos of rebirth.

Rather than purposefully seeking wilderness's solitude to rejuvenate one's self as Fletcher, Abbey, and so many others do, McCairen's description of this "rebirth" suggests that her experience was entirely unexpected. She explains that she had no experience on rivers or camping in the wild—her narrative suggests that at this point she was entirely ignorant of the so-called promises of the wilderness myth; she had no expectations for what would happen on the trip. And unlike Fletcher's desire to regain a sense of purpose and his identity as a man, McCairen's experience suggests more of a first birth rather than a rebirth. Having tried to conform her life according to certain gendered identities dictated by society's standards for beauty and how a woman should behave for men, Babe's appearance not only gives McCairen the first real sense of who she is, but also challenges the traditional woman-as-temptress character that defines many westerns.[11] This reformulation of the topos of rebirth in wilderness takes on a new meaning as McCairen advocates for her new beginning. Although McCairen replicates a journey down the Colorado similar to so many men before her, she reorients the voyage to speak to the freedom that comes not from fleeing civilization because of its effeminizing effects, but from casting off societal expectations regarding her gendered identity that have prevented her from seeing a more complex and honest version of herself.

The emphasis McCairen places on her rebirth as Babe, a woman whose identity transcends limiting and oppressive views of women, is further reinforced through her passage through the rapid. While her descent into the Inner Gorge, the Grand Canyon's deepest and narrowest stretch, is a clear metaphor for her entrance into the womb, her journey through the rapid is depicted not through exploitative, sexualized terms between an adventurer conquering the river as in Abbey's account, but through the maternal experience of giving birth. Similar to Hyde's shift in perspective discussed in this chapter's introduction, McCairen attempts in this scene to construct the river and canyon as a woman's place, distancing it from the typical feminization and sexualization of the river by various male authors.

While this scene initiates McCairen's journey to establish her own rightful position on the river, it is in her actual solo voyage down the river and her reflections on her past that she further challenges the exclusivity of the Colorado boating experience and reveals the ongoing process of her rebirth. Central to achieving her new life is a reconciliation with the feelings of inadequacy she feels as a woman and the fear of being alone. In planning her Grand Canyon solo trip, she confides that while motivated to rekindle the sense of freedom of empowerment experienced at Babe's birth, and while working as a guide on smaller rivers throughout the West, she was gripped by the fear of solitude. Her chapter "Going Solo" charts her apprehension, recognizing that her real fear was less of what could happen but "the aloneness of it. Facing whatever there was to face alone" (39). While a solo voyage down the Colorado is no small undertaking and certainly worthy of such concerns, McCairen's worries are all the more exacerbated by the ongoing lack of confidence she recounts from her experiences as a woman. She recalls attempts to secure employment as a Colorado River guide and the subsequent rejection because, "though I looked and felt ten years younger, I suspected that at forty they thought I was too old, though men my age had no problem being hired" (70). McCairen explores the sting of this rejection throughout the text, questioning it later on when she meets another rafting party at one of the river's many sandbars. Seeing that most of those in the group are men, she explains that she desires to show "how brave and independent I am" in an attempt to prove to herself and them that she belongs on the river (132).

McCairen's preoccupation to justify who she is and to give Babe full expression leads her to look to the river for answers and guidance. As the river first gave voice to Babe, it now becomes her mentor, leading her out of the depths of insecurity to see a bright future where she takes charge of her own life. By studying the water and learning the techniques to negotiate the river's obstacles successfully and safely, she discovers that "using the river's strength, direction, and flow helped me more than absolute control or total passivity" (170). This ongoing interaction with the river over a number of years facilitates this ongoing awakening that McCairen experiences and establishes a close connection with the river that one should not attempt to completely control or give way to. She speaks to an intimate balance of give-and-take that allows her to move safely downriver and that provides a useful metaphor for our own considerations of how we interact with the river.

McCairen continues to rely on this river-as-life metaphor as a way to chart her rebirth. In her conclusion, she reflects back on the lessons learned while on the river and acknowledges, "Some days I'll have perfect runs, and other days I'll eddy out and flounder around in murky water before continuing on downstream. But unlike in the past, now *I'm* the one at the oars" (246). Until her solo voyage, McCairen lived her life according to the dictates of others. But faced with challenges that only she can overcome, the journey down the Colorado allows her to discover her strength of character that was dormant until given a chance to grow. Far different from Fletcher's and Abbey's narratives that seek a revitalization of what they see as their true selves once the facade of civilization is removed, McCairen's account of her rejuvenation is a discovery of, not a return to, her true identity. As she heads off into the wilds of the canyon, she overturns the wilderness myth's flight-from-the-city topos to suggest that it is just as much the oppression of a male-dominated society that one needs to escape from as the effeminizing effects many male writers attempt to relinquish in their own quests for the wild.

Associated with the topos of rebirth is McCairen's negotiation of the domestic sphere through her reflections on the relationship she has with her mother. Described as a hypercritical woman, McCairen's mother represents the conventionality that the author seeks to escape. As she reflects on her journey that led her to a solo voyage through the Grand Canyon, McCairen notes the oppressive influences she felt dictated her life and looks to her mother as the full expression of that oppression. McCairen writes of a "society that favored boys over girls" and the many male teachers throughout her adolescent years who "discouraged girls from contemplating college or pursuing a career" (93, 94). Yet she explains that in respect to this latter group, "My mother was worse." McCairen elaborates, "Whether I expressed interest in acceptable female professions such as teaching or library science, or traditional male occupations such as zoology or law, her response was the same: 'College will be a waste. You're just going to get married and have children'" (94). Negotiating her mother's expectations with the unhappiness McCairen witnessed in her parents' marriage, McCairen concludes, "Getting married and having children was a fate worse than death" (188). With these new realizations, McCairen pursues a life that on the surface rejects this domestic future as she seeks out solitude and self-discovery in the West's mountains and canyons.

However, while McCairen brusquely turns her back on the life her mother anticipated for her, she nonetheless embraces the elements of the home as she rethinks her place within the canyon. When she is reborn in the rapid during her first visit to the canyon, she speaks of "My Mother River, My Father Canyon" (20) and her alter ego, Babe, who "like a newborn screaming in her crib...demanded nourishment and attention (21). By evoking the nuclear family, McCairen's text reveals a latent desire to have the security and support represented within the ideal of the traditional home and family. Yet she constructs this home on her own terms, setting up as parental figures the river and canyon to which she now attaches herself. When she passes safely through Crystal Rapid, one of the most feared sections of white water on the Colorado, McCairen remarks, "This river is my mother and she loves me" (171). Recognizing the failings of her own family life shaped by an absent father and a bitter, depressed mother, McCairen reconstructs her lineage within the canyon to formulate her ideal version of the family based on love, openness, and freedom.

Just as McCairen recasts the domestic sphere to better align with her own desires, she questions what she feels are unequal and unfair expectations that define men's and women's behavior and that prevent women from seeking adventures like her own. "Men, for the most part, have not been labeled peculiar when they go adventuring alone," she writes. "Rather, they are considered brave and daring. Of course, by keeping women tied to the kitchen and bedroom, a man has someone to come home to, someone to swoon over his heroic deeds" (199). Reacting against these constraints, McCairen posits an alternative life for those like herself who are independent and prone to wandering: "There are some of us who simply can't be tied down. It's a positive response to our nature rather than a negative reaction to our past. My restlessness and need for change, my curiosity to see the world, may make me an unsuitable mate" (200). McCairen's acknowledgment that her life is probably better off by being alone certainly hints at the male writers' desire for solitude in far-off places. Yet her consideration of a life of solitude in and out of wilderness carries this desire beyond what the men's narratives suggest. Rather than a periodic escape from society to clear one's mind, McCairen's work suggests a much more comprehensive withdrawal from society, domesticity, and the traditional woman's life. She must rely on the topos of solitary adventure typically associated with men in these narratives in order to distance herself from the conventional role of women, but as she extends it further than most male river narratives do, McCairen draws up a far more complex and conflicted perspective of what it means to live between these competing narratives that position adventure and freedom within a man's world and confinement and disappointment within a woman's.

McCairen does not clearly resolve this conflict within the text. In the final pages of the narrative, her quest for freedom through solitude turns on its head. As she moves slowly through Lake Mead's impounded waters at her journey's end, she realizes that her "search for freedom has led me into a trap. It is only through commitment—first to myself then perhaps to another—that I can ever know complete freedom" (245). For McCairen, this solo journey through the depths of the Grand Canyon has allowed her time and space to probe the recesses of her soul to better understand her responsibility to herself and others and the liberty that can come by attending to both. Recognizing that she does not have to choose a life of complete isolation or of compromise, she reaches the trip's terminus enlightened by the emotional and physical strength honed on the river. Realizations such as these place *Canyon Solitude* in line with countless other texts in American literature where Groover's articulation of the spiritual quest is at the heart of the work. Nonetheless, McCairen's ability to work within and against typical conventions of the androcentric river narrative embodies the complexity that McCairen and other female adventurers like Teal's guides must face as they describe their presence on the river. Torn between expectations of whom and where they should be, they negotiate conventionality to create a space for their own distinct voice.

Buyer's *Side Canyons*

Similar to the previous two texts, Laurie Buyer's *Side Canyons* also engages the western and its poetic discourse to reimagine the spiritual-quest narrative and a woman's relationship to those around her and the land. Yet unlike the others that craft their accounts through nonfiction, *Side Canyons* disrupts the genre conventions of the traditional Colorado River western by blending poetry with the author's fictitious account of her own real-life journey through the canyon in unpredictable ways. This method is evident in the opening section of prose, "On the Ranch," which Buyer punctuates with poems "At Wellington Lake," "Disfigured," "Live Analysis," "Weeping," "Portrait of a Woman in a Box," and "Interlocking Limbs." In a formal sense, this experimentation departs from the linearity and predictability that typify river narratives. While Buyer's text does proceed chronologically overall, the poems move in different directions temporally and spatially as they shift between the narrator's interior and exterior experiences. That is, the poems inject a greater awareness of the narrator's feelings that are less developed in the prose. Like Teal, Buyer expands the genre choices for women boaters by which to articulate their unique connections to the canyon and river, thereby boldly signaling her departure from the status quo.

The text centers on Laurie, a woman depressed over a failing marriage with a man who is emotionally withdrawn except when drunk. Seeking respite from this relationship, she agrees to join Angie and her group of female friends for a vacation down the river—a journey that becomes Laurie's own spiritual quest. Like so many other Colorado River westerns by men and women, *Side Canyons* emphasizes the wild force of the river, the sublimity of the canyon, and the peace found in solitude. And similar to McCairen's account, Buyer's narrative does not wholly reject civilization but works to acknowledge the role that a shared community plays in mending a broken heart and soul. Worn out by years of arguing and falling out of love, Laurie moves between needing her own space and time to her own thoughts while on the river and the support and reassurance of her fellow rafters. Although she is accused of "pick[ing] the oddest places to camp" as she chooses secluded areas where she can commune with her surroundings, she also welcomes the company of others and has no problem joining a friend for a walk along one of the canyon's many beaches (92). Laurie's openness to her fellow travelers in a place where so many Colorado River westerns emphasize isolation gives nuance to Groover's conventional notions of community and reinforces its significance beyond Teal's and McCairen's accounts. As Laurie welcomes a community of like-minded river travelers, she turns her back on the domestic community from which she and so many of her sister boaters escape, even if only temporarily.

This more egalitarian community is evident later in the text when Laurie finds herself reflecting on the unique relationships she has established during the trip after a late-night hike up a side canyon with friends. She explains:

> I pulled my flannel bag out into the moonlight and lay there thinking about all the hands I'd held in the Canyon, how we helped one another get up from our soft seats in the sand or assisted each other in and out of the boats. While hiking, outstretched hands reached to pull me up a difficult place or guide me down a tough spot and a chain of hands guided me across hard-current creeks.... I never reached out my hand that someone, man or woman, didn't take it, hold it, give it a quick squeeze, or pull into an embrace. In this unknown environment, holding someone's hand meant a moment of sheltered harbor where I felt safe. (185)

Contrasting with the emotional and physical distance in her marriage, the repeated contact of another's hand sustains Laurie throughout her voyage. When faced with an obstacle, she finds reliable support as her fellow travelers cooperate to lead her safely onward. Linked with this unique community bound together by a shared interest in the place, Laurie's reflections on these relationships offer a unique way of thinking about the river and canyon outside of the typical "isolation brings healing" topos. Just as McCairen is unable to entirely disregard community, Laurie sees the necessity for communal relations even in a relatively isolated locale.

Buyer also plays with this theme through the role that communication plays in the text. Laurie, a writer, is passionate about her craft, having traveled throughout the country to attend various workshops. It is through this intense desire for expression that she feels herself growing apart from her husband, John. Buyer portrays John as the traditional western man: He's a tough Wyoming rancher of few words.[12] When Laurie first broaches the subject of going on a raft trip through the Grand Canyon, John is concerned because she will be gone when the calves need to be shipped. When Laurie asks whether he could wait until she returns, John retorts, "Forget it. I can do it alone." And when she suggests that he should ask the neighbors for help, Buyer notes how "he just stared at me with unspoken reproach." When John does speak, he questions why Laurie has been away so much recently, adding the raft trip to her attendance at a writing workshop and enrollment in a Vermont school that have taken her from the ranch for extended periods of time. Interestingly, Buyer recounts that as John says all of this, he is peering out the window at their dog, Blue. Rather than looking at Laurie in an attempt to read her expressions and understand why she wants to leave, he stares at man's best friend. While a small detail, it underscores what Buyer attempts to communicate through her depiction of John. With his attention directed to his dog and his reluctance to respond to Laurie, John isolates himself from his wife's attempts to engage him and thereby rejects this moment to forge an emotional connection. Unwilling to face the reality of their situation that Laurie is trying to communicate, John finally asks, "Well, I wish I knew what the hell it is you're searching for" and then leaves the room. Buyer closes the scene with Laurie responding but to no avail. "Not knowing if his nonresponse meant that he had not heard me or that he chose to ignore me," she concludes, "I did not follow" (18). John flees the situation, perhaps going to consult with his dog, a stand-in for the western hero's faithful steed.

Contrast Buyer's characterization of John and his complete aversion to communicate and nurture relationships with Robb, the raft trip's lead guide who, in many cases, preserves the image of the classic western hero. Buyer reinvents the male river guide as she casts Robb as a more sensitive man, one who welcomes community and is willing to listen and express feelings while retaining his charisma, knowledge, and masculinity. Instead of creating a guide like Powell, whose indomitable, larger-than-life ambitions often created tension within the expedition, Buyer transforms the convention to depict her leading man with a greater sense of compassion and acknowledgment of others' needs—qualities John seems incapable of cultivating. Throughout the journey, Laurie gravitates to Robb, admiring his humility, kindness, and reverence for his surroundings. On their third day on the river, just below Spector Rapid, Laurie considers her interest in Robb. She describes him as "a keen observer, not only of the Canyon and the river, but also of all of us in his care. Calm and quiet, he appeared everywhere at once." She continues: "He pitched in with every facet of camp life: he cooked and washed dishes, moved the [toilet], loaded and unloaded the gear, built fires, helped set up the kitchen and take it down again. . . . [H]e remained patient and kind, spending time with each person in camp. Attentive to everyone's needs, he seemed to know before anyone asked what was wanted" (107). Like some river god, Robb is both omniscient and omnipresent, fully in control of his environs, never bothered by any situation no matter how big or small it may be, never afraid to work. He's also willing to perform tasks traditionally assigned to women.

And these qualities cause Laurie and her friend Lana to express their deep admiration for the man, suggesting that they too could possess "his connection to the Canyon, his affinity for wildness, this love affair he has with the river" (109). Through such comments, it is clear that Robb is the object of these women's desire, not so much in romantic terms, although there are repeated hints of this throughout the text, but as the embodiment of everything they wish the men in their lives could be: free, calm, at perfect ease with oneself and the world. Of course, one expects a guide to attend to a paying customer's needs. However, the idealism and earnestness with which Buyer endows Robb set him apart from even the most attentive guides.

While Robb embodies many of the qualities of the typical male hero, his depiction questions the text's success in countering traditional views of women. "Sharing an easy camaraderie with the men," Buyer writes, "he offered his knowledge of the river and the Canyon openly. Sensitive and conciliatory with the women, he offered a sympathetic ear and a tender touch. Never playing favorites, he found ways to be magnanimous to each of us" (108). Robb does seem larger-than-life. He is both strong yet sensitive—he is what a "real man" should be, Buyer's ideal. As such, Robb communicates with the women and men on the terms Buyer suggests, which, despite efforts to show how Robb is different from John, also reinforce traditional gender expectations. Suggesting that Robb uses reason to engage the men and emotion to appeal to the women, Buyer somewhat undermines the ground she has gained in challenging norms. However, on the following page, she is quick to clarify that knowledge is also shared with the women, as "he tells us about astronomy, geology, and biology, the environment" (109). Herein lies the difficulty of trying to make Robb everything to everyone. To challenge the male hero, Buyer elects to make him more in tune with his female companions, but in doing so she establishes clear divisions between the other boaters who represent traditional female and male roles.

Notwithstanding this example, the narrative successfully critiques the western's depiction of the hero by reaffirming Robb's penchant for communication that meets the protagonist's needs. When Laurie and company near the end of the trip, she is gripped with melancholy, sad to see the journey conclude when she has seen and experienced so much new physical and emotional territory. After some misunderstandings between the women come to a head, Laurie shrinks off to find Robb, in whom she can confide. She explains, "There's so much I want to say," yet she struggles to express all the feelings inside her, feelings of hurt, sadness, love, and liberation. When Robb replies that he understands, Laurie wonders "if he could know that the river changed me in significant ways, that I felt like I finally knew who I was, that some long-ago wound had healed" (240). The river experience proves in many ways to be the balm Laurie so desperately needs to heal her soul, and such a transformational experience for this woman would initially seem difficult for a man of John's type to understand. But not for Robb. "Yes," he answers, "I do. I understand the words you don't know how to say" (241). Like Teal, who learns to understand the river's story without words, Robb reads Laurie to sympathize with her newfound sense of self.

Interestingly, despite the numerous instances where Laurie emphasizes the need for companionship in the canyon, she explains that she probably would not want to return with the same group on another trip. She explains to Angie, "There are just too many people. It's too hectic and unsettled for me. If I come again I'd like to find a way to go to a remote place and be alone with the Canyon" (252). This statement seems to undermine all that Laurie has said about her need for others. However, in light of the transformation that has taken place within her, she is now able to face challenges more on her own two feet. Such a statement is not possible before this trip, as she was too vulnerable, too insecure. But having drawn strength and experience from this small boating community, she is now confident to navigate her own path in the canyon on a more intimate and independent level. Asked whether the river trip changed her life, she replies, "Yes. I'm not the same. I don't know what that means, but there it is" (251). Empowered by the community experience, Laurie must now decide her future. While much self-discovery awaits her, she is now endowed with greater confidence and buoyed up by the healing effects of the raft trip to face whatever challenges lie ahead.

The initial pain and suffering that brought Laurie to the river in the first place have given way to a new independence and resolve. But this liberation does not come from fleeing society or imposing one's will on nature, as the traditional western and Colorado boating narrative often suggest. Rather, it evolves from working with others and listening to what the canyon has to teach, just as Teal and McCairen demonstrate. On her voyage Laurie articulates the lessons she has learned from observing her experiences: "Go slowly. Step softly. Take time to look and listen and feel.

Be considerate. Help out. Hold hands. Allow yourself to be immersed in awe. Make reverence a daily ritual" (150). Although a Romantic strain runs through this passage with its attention to the aesthetic and spiritual sublime, Buyer places the transformative experience alongside the companionship of others. With this revelation, Laurie can return to John and her Wyoming ranch equipped with the tools not only to survive but to thrive. With a new sense of self born on the river and forged through the relationships she made with her traveling companions, Laurie may embrace her future with determination and a contentment that the memories of being on the river will always be there to remind her of the progress she made in the company of others. This subtle reorientation of the wilderness experience embraces a different set of values that reimagines the genre as something much more than a tribute to the Colorado boating experience. Emphasizing the role of community throughout the text and the role it plays in empowering a woman to face with confidence future trials, *Side Canyons* advocates for a reconsideration of the river experience and what benefits can come through community to those lacking a voice or the power to create such change.

Ryan's *Writing Down the River*

I conclude this analysis of women's Grand Canyon boating narratives with Kathleen Jo Ryan's *Writing Down the River,* as it offers a more nuanced view of the women's Grand Canyon boating experience than those previously addressed. In her introduction to the collection, Ryan explains that this book came about as a result of an invitation to join a friend and river guide on a Colorado River rafting trip. Having just completed a book about ranching traditions (of the same name), she felt that doing a "river trip would be fun, a change of pace from traversing the American West by car and on horseback" (xxi–xxiii). In sum, she swapped "bucking broncs" for "wrestl[ing] rafts," as McCairen writes (10). Transformed by this experience on the river, Ryan searched "for a book to validate and celebrate my feelings for the Grand Canyon and Colorado River experience. I did not find such a book" (xxiii). Since most texts were by men and those by women were exclusively by white women, Ryan sought a more inclusive project that extended the community of river rafters very much absent from the tens of thousands who float the canyon each year. Along with short selections by Annick Smith and Ann Zwinger, for example, the collection of sixteen women writers includes brief entries by Native,

African American, and Chicana authors such as Linda Hogan, Barbara Earl Thomas, Evelyn White, and Denise Chávez. Like the women in Teal's, McCairen's, and Buyer's texts, those represented in *Writing* also rely on the discourse of wilderness to conflate the public and private, but they do so with a greater awareness of how race and class shape one's time on and off the river.

Consider the experience of Evelyn White as she awakens early one morning with Tina Turner's "Proud Mary" playing repeatedly in her mind: "I offer a quiet prayer of gratitude to my African ancestors. I thank them for bringing me, the only black person in our group, the spirit of Tina Turner: a free and talented black who rose triumphantly out of an abusive marriage to strut her stuff" (29). For White, the trip to the river represents her attempt to overcome her fears of being in nature. Along the way she comes to recognize how one's understanding of and experience in wilderness are mediated by race. "I didn't realize the full significance of my journey," she explains, "until I found myself explaining, to the river guides, the complex, psychological role water has played in the lives of most African-Americans." As she recounts to her group of guides, who all happen to be white women, the traumatic cultural memories of the Middle Passage, segregation laws, and firehoses unleashed on civil rights protesters, she watches "their stunned amazement," their "acknowledgment of a never-considered truth" (30). Meeting a black man at Havasu Falls gives further insight into this revelation: "All the lessons we are taught in our families, church, and community are about moving forward and getting ahead.... To come to the wilderness is to return to the primitive. Black people don't see anything 'advanced' about sleeping outside or relieving themselves in the woods" (33). This perspective is reiterated by Barbara Earl Thomas as she reflects on her journey down the river: "It sounds so ridiculous that I hesitate to say it out loud, but I don't think this is something a black person is supposed to be doing. Even the old bus driver, the only other black person I've seen anywhere near here, drops his jaw when he sees that I am part of the group" (93). Yet for both White and Thomas, the river trip is one of emancipation from their own fears and discomfort.

Like White's and Thomas's experiences, Denise Chávez's contribution is also mediated by race as she declares, "I may be the only Mexican rafting the Colorado." She then asks her guides if other "ethnic people" do the canyon trip. "Few, they both say. I wonder why? Is it the cost? It *is* prohibitive, but not for all. It's more than that" (118). She goes on to explain, "We are a people who want comfort, manageable thrills, self-made happiness. Our blood runs too deeply down the muddy river. We were and are its slaves, its laborers, its workers, its drones. The memory of hardship is still too much to bear. Our ancestors tried to conquer this river, but they turned back too soon. On land the Spanish ruled, but in this hallowed place they met their match" (119). Chávez's observations note the historical and economic connections the Colorado River has to her people, La Raza, as this portion of her account is titled. The canyon and river are not blank slates, a nostalgic return to Eden, or a "silt-brown bosom" in Abbey's case but a place of lived experience and deep memories reflecting the promises and heartaches of living in an arid land. Though brief, Chávez's essay as well as those by White and Thomas directly challenge the wilderness and frontier myths so steeped in Anglo discourse and experience.[13] Rather, these women's presence on the river and reckoning with historic and ongoing racism dramatically shift the river into the New West, where tributary voices enable the river's "whole story," as Teal puts it, to rise to the surface.

In "Wild Women: Literary Explorations of American Landscapes," Sarah McFarland considers the absence of women from an American nature-writing tradition that "has been burdened by a discriminatory history, influenced by wilderness exploration and scientific study that excluded women." Like Groover, who challenges this forced absence in the spiritual quest, McFarland argues, "The exclusion of women from nature writing is not a problem only because it disguises the fact that the very concept of wilderness is a male construct, but also because it leaves careful readers with the impression that nature is no place for women" (41). While the accounts I have considered in this chapter do not follow the typical conventions of nature writing, with its botanical and zoological considerations as witnessed in the previous chapter's considerations of desert explorers and adventurers, these women's narratives nonetheless demonstrate that nature, and particularly the Colorado River and Grand Canyon, are very much a place for women. As these women travel through the canyon, learning to navigate the river's many challenges while drawing on their surroundings to make powerful connections to their own lives, they are empowered as they shed prejudices and find a new sense of individual worth.

The lessons that emerge in their texts as we read them in light of the western genre demonstrate why such a consideration is particularly valuable to understanding how the Colorado has shaped both the imagination and the actual lives of those who have traveled its renowned waters. While westerns have often been disparaged for their perpetuation of stereotypes and escapism, they prove useful as they reveal much about what a culture values and what it denigrates. In the case of the Colorado River westerns of the Old West, we find that the river is a place for men to find adventure, beauty, fame, and a panacea from the ills of urban living. By examining these women's texts as Colorado River westerns that appropriate the genre's principal themes for their own purposes, we come away with a very different picture of the river and for whom it flows. More pointedly, as we consider the texts that celebrate the beauty and wonder of the river and this canyon, we see how they appropriate a genre and its many topoi to create what then become powerful rhetorical tools that create room for marginalized perspectives. Jane Tompkins reminds us in *West of Everything* that the western has "influenced people's beliefs about the way that things are." If, as Tompkins further suggests, "what the hero experiences is what the audience experiences; what he does, they do too," then it stands that alternative models for western heroes could have equal sway in shaping the public's perception and actions about the river and region (6). By reading the female boating narrative into the traditional and historical discourse of river accounts, we open ourselves to stories about the Colorado emerging from the New West, stories like those evident in these women's texts that envision a new reality for the river. With a watershed under such distress by the endless competing demands of government, commercial, and private interests, such a shift in perception from "the way things are" is needed now more than ever.

CHAPTER FOUR

Nuestro Río, the Goodlife, and *Querencia*

Hispano, Latina/o, and Mexican Perspectives on the Colorado River

El Agua no si vende, si defiende
(You don't sell water, you defend it)
—Albuquerque bumper sticker

On September 30, 1935, President Franklin Roosevelt addressed the crowd at the newly completed Boulder Dam, triumphantly exclaiming, "This morning I came, I saw and I was conquered, as everyone would be who sees for the first time this great feat of mankind." With these words, Roosevelt opened a new era along the Colorado River dominated by an ardent desire to reclaim the arid West through human industry and ingenuity. In doing so, he further cemented the river and surrounding desert as worthy opponents to overcome. He proclaims, "Ten years ago the place where we are gathered was an unpeopled, forbidding desert. In the bottom of a gloomy canyon, whose precipitous walls rose to a height of more than a thousand feet, flowed a turbulent, dangerous river. The mountains on either side of the canyon were difficult of access with neither road nor trail, and their rocks were protected by neither trees nor grass from the blazing heat of the sun. The site of Boulder City was a cactus-covered waste. The transformation wrought here in these years is a twentieth-century marvel" ("Address at the Dedication"). This engineering feat became the envy of the world as the ultimate symbol of progress—a fitting monument to the realization of the garden and frontier impulses and humans' conquest over nature. Closer to home, Boulder Dam also inspired the Upper Basin states to consider how similar reclamation projects could help them develop their own share of the Colorado River allocated to them in the 1922 compact.

In 1956 Congress passed the Colorado River Storage Project Act (CRSPA), a key piece of legislation within the Law of the River, that body of laws, policies, and regulations that govern Colorado River management. The act approved the construction of four major reclamation units by the Bureau of Reclamation throughout Colorado, New Mexico, Utah, and Wyoming on the mainstem of the Colorado River and a few of its primary tributaries, as well as a number of smaller "participating projects" to further enhance the region's water infrastructure ("Colorado River Storage Project"). Over the next decade, the bureau constructed major dams, including Glen Canyon on the Colorado, Navajo on the San Juan, Flaming Gorge on the Green, and Blue Mesa on the Gunnison. The fiery bureau commissioner Floyd Dominy referred to Glen Canyon, the second-largest dam on the Colorado after Boulder, and its massive reservoir, Lake Powell, as the "Jewel of the Colorado" and "the finest scenic and recreational area in the Nation" (126). Such enthusiasm for dams and their promises for economic prosperity reflected the nation's priorities in the post–World War II era.

However, not everyone embraced the Colorado's development as did the region's politicians and chambers of commerce. While Boulder (later Hoover) Dam became a symbol of America's ability to pull itself out of the Great Depression, Glen Canyon Dam has elicited a far more checkered response. For those seeking to follow in Powell's footsteps and experience a "pure" and "wild" river, Glen Canyon became a symbol of the river's demise—a direct affront to the wilderness experience pursued by many of the men and women discussed in the previous chapters. When Glen Canyon's floodgates closed on March 13, 1963, the dam inundated, in Edward Abbey's words, "a portion of the earth's original paradise" (189). Many of the region's most notable adventurers, literary figures, and activists have joined Abbey's lament, creating a robust body of discourse labeled by historian Jared Farmer as the "literature of the lost" (xiv). Georgie Clark, the pioneering river runner, describes Glen Canyon Dam as "a travesty for which we shall all pay forever" ("Conquering" 102). It and other dams "destroyed the entire Colorado adventure as I used to know it,"

Clark recalls, "and have changed the entire complexion of river running everywhere, forever" (111). Katie Lee, one of the most outspoken critics of the dam, makes no qualms about how she feels about the dam and its impact on the Colorado: "I can't go crazy here like I did in Glen. In spite of being alone, this whole show is less electrifying, less profound.... There hasn't been a Little Arch, a Dove, a Grotto, Driftwood, Twilight, or Cathedral among them. I'm thinking 'soiled dove' for this place. Most of what can happen to me here is predictable. Mystery has been submerged beneath the rising tide of population. It's no longer virgin territory, nor was it even then, when the Glen was truly pristine; when we had the unparalleled good fortune to embrace its secret places...years before the rape" ("Split Bar" 12). For other writers, just the awareness of Glen's losses is enough to evoke anguish. Although Ellen Meloy never experienced the canyon before it disappeared under Lake Powell's rising waters, she nonetheless shares Clark's and Lee's pain as she observes the reservoir and imagines what lies beneath. In *Raven's Exile* (1994), Meloy confides, "Lake Powell's bathtub ring, sixty feet high in this interminable drought. Suddenly I am weeping. I cry because I cannot smell the heady musk of cottonwoods or feel the quiet movements of water in the unlikely ribbon of paradise that curls through acres and acres of bare, baking slickrock. I cry because 3 million people a year love the Blue Death and I don't. I weep for Glen Canyon and I have never seen it" (94). For these women, a deep sense of betrayal exists as the indescribable beauty and tranquility of this desert canyon—this paradise and "virgin territory"—have been ravaged by humanity's insatiable desire for growth and progress.

These women's powerful sense of loss embodies what many have viewed as one of the nation's most grievous environmental sins. As writer and adventurer Wade Davis claims, Glen Canyon Dam "is widely perceived by the elders of the American environmental movement as one of the most egregious acts of betrayal in the history of the country" (23), while Derek Ross suggests that "the dam has become a focal point in relation to modern environmentalism" (77). Perhaps most significantly, David Brower, the head of the Sierra Club who led the charge against Glen Canyon Dam in the 1950s and '60s, famously admitted that the dam was "America's most regretted environmental mistake" (qtd. in Grace 111).[1] For those committed to America's wilderness ethic and the anti-dam legacy that emerged from John Muir's protest of Hetch Hetchy Valley's damming in Yosemite National Park in the early 1900s, Glen Canyon Dam represents our society's failure to preserve and value beauty, silence, and solitude.

Yet Glen Canyon Dam's long-held position as the poster child for the demise of American wilderness is far more complicated and problematic than most people familiar with Glen's history assume. As this chapter argues, by continuing to position Glen Canyon as "The Place No One Knew," to borrow from the title of Eliot Porter's photo book about Glen Canyon's many natural treasures, we favor the oft-cited exploration and wilderness accounts such as those referenced in chapter 2 rather than those narratives from the Colorado's tributaries dammed by CRSPA projects. That is, by focusing so intently on the environmental losses in Glen Canyon, one fails to notice the presence of those whose homelands and livelihoods have been directly impacted by dam construction throughout the watershed. Similarly, such preoccupation with Glen Canyon overshadows how perspectives voiced by those including Denise Chávez, Linda Hogan, Barbara Earl Thomas, and Evelyn White from chapter 3 reflect larger issues about how minorities experience the river today in a culture dominated by leisure and the desire to "get away from it all."

Thus, this and the subsequent chapter critique America's love of wilderness and the rhetoric promoting an unpeopled West by examining more closely how dams and other water infrastructure throughout the watershed have impacted Hispano, Latina/o, Mexican, and American Indian communities.[2] Simply, they have been the victims of what Rob Nixon identifies as "slow violence," defined as "violence that occurs gradually and out of sight, a violence of delayed destruction that is dispersed across time and space, an attritional violence that is typically not viewed as violence at all" (2). Historically marginalized within the West's water order, these tributary voices constitute a new literature of the lost that is nonetheless integral to the region's development over the past four centuries and powerfully remind us that the basin never was a blank slate or arid Eden that "no one knew." Rather, the basin's rivers have supported vibrant communities whose connections to place rebuff society's rejections of culture in favor of the wilderness experience and its associated values of leisure, solitude, and the individual. Regarding this relationship between marginalized groups working to preserve their connections to place and such values that influence much environmental thinking and action, I build on the editors' assertion in *Latinx Environmentalisms* (2019) that "the whiteness of mainstream environmentalism fails to account for the diverse environmental ethics in communities of color, including Latinx cultures" and that these "cultures redefine and broaden what counts as environmentalism" (Wald et al. 3). In this chapter, I account for this range of environmental ethics by examining a variety of textual representations by Hispano, Latina/o, and Mexican voices and their respective stories about the river to then foreground their alternative perspectives and vocabularies reflective of their unique discursive ecology of water knowledge for living with the Colorado and its tributaries. Situated within the backdrop of Navajo Dam in northern New Mexico and the Colorado River delta in Mexico, these discourses and the values they convey foreground environmental justice issues related to dam construction that have historically existed and continue to persist along the river today.

"Slow Violence" and Navajo Dam

Nixon's influential work on environmental injustice and its representative examples of injustices afflicting the world's poor communities provides an important springboard for this chapter's discussion of Hispano, Latina/o, and Mexican perspectives on the Colorado River. Interested in the impact of megadams across the globe among other issues, Nixon considers their role in the Global South, in particular, to identify how these massive structures create "spatial amnesia, as communities, under the banner of development, are physically unsettled and imaginatively removed, evacuated from place and time and thus uncoupled from the idea of both a national future and a national memory" (151). In the case of his reading of India's displaced poor, he identifies them as the "unimagined communities," the "uninhabitants," and the "developmental refugees" that fall to the wayside of India's narrative of nation building (152). Juxtaposing India's anti-dam movement with America's, he looks to Muir, Abbey, Brower, and Stegner to conclude that their resistance to dams emanates from "traditions of the wilderness ethic," which "borrows from and advances the rhetoric of an international environmental justice movement that did not exist in any comparable form during the years when [Abbey], much less [Muir], was writing." To further draw this distinction between these case studies, Nixon suggests that India's dams were constructed in "rural areas densely populated with subsistence farmers, areas quite unlike the thinly peopled hinterland of the American West" (155).

While these conclusions draw important distinctions between India's and the United States' anti-dam campaigns and their influence on the rise of environmental justice initiatives, this reading of America's anti-dam legacy leads one to believe that America's dam building destroyed only botanical, zoological, and geological treasures. Therefore, as Nixon locates America's anti-dam literature within this aesthetic, he unintentionally reinforces timeworn tropes about the arid West that replicate the very erasure he seeks to critique. Put another way, if we continue to forward a West-as-wilderness narrative by overlooking the displacement of Hispano and Indigenous communities in the American West, which speaks to the slow violence they have experienced and the unique discourses that preserve their community identity, we position such groups in a manner that they "can be readily dismissed as living benighted lives," relegated "in the fullest sense of the dead metaphors, to a cultural backwater not the national mainstream" (164).[3] In reading the Colorado River Basin through the lens of environmental justice rather than the wilderness ethic so readily apparent in Glen Canyon Dam's many responses, the region populates with voices to create what Devon Peña identifies as "terrains of struggle" or the sites in which Latina/o environmental justice issues arise and are expressed ("Scope" 48).

There are many advantages to reading Navajo Dam and the Mexican delta in this manner. First, we avoid the pitfall of merely "calling Western, non-dominant works 'environmental,'" as this "holds the potential to ignore the intersections of social justice, identity, national belonging, and cultural tradition that shape different communities' environmentalisms" (Wald et al. 9). Instead, framing them as terrains of struggle acknowledges the region's history of environmental racism toward Hispano, Latina/o, Mexican, and Indigenous communities as it pertains to river management, as well as the complex relationships these groups have with one another. Also, it reorients our vision from the pervasive wilderness discourse that infuses many conversations about dams in the US West to embrace a broader kind of environmentalism that respects the presence of local communities and their respective connections to their surroundings. Finally, this reorientation provides alternative vocabularies such as the goodlife and *querencia* by which to reframe discourse about relationships to the Colorado and its stakeholders.

Approved as part of the CRSPA and completed in 1963, the Navajo Unit comprises Navajo Dam and Lake. Located in the high desert just south of the Colorado–New Mexico border on the San Juan River, Navajo Dam draws a stark contrast from Glen Canyon located downriver. Unlike the massive concrete structure that stands 710 feet tall at its crest and wedged between towering sandstone cliffs, Navajo Dam is a 402-foot-high earthen dam whose reservoir impounds a little more than 1.7 million acre-feet ("Navajo Dam"). By contrast, Lake Powell holds a sprawling 27 million acre-feet and stretches 186 miles upriver when at full pool ("Glen Canyon Dam"). These differences, and the fact that Navajo Dam's construction occurred with much less controversy than Glen Canyon's, likely describe why Navajo has long been in the shadow of its more notable neighbor. However, by shifting our attention to the events surrounding Navajo's construction and its impact on the local communities, this lesser-known dam becomes a more appropriate symbol for the American environmental movement than Glen Canyon.

As Katie Lee wisely acknowledges in her lament of Glen Canyon's demise, it was no virgin wilderness, even though its remote location could easily lead one to believe that it was untouched by human intervention. In fact, the Ancestral Puebloans had long used the canyon to store food in granaries that dotted the canyon walls and establish lookouts or temporary shelters, but since 1300 when these peoples moved farther south along the Rio Grande, no permanent habitation existed in the canyon (Martin 111). Thus, Glen has often been considered more of a place to pass through than to settle, despite the canyon's continued significance to many of the area's tribes, who have cultural and spiritual connections to the canyon.[4] On the other hand, Navajo Dam's construction history presents a more immediate and direct impact on the livelihoods of those inhabiting lands eventually flooded by Navajo Lake. Constructed below the confluence of the San Juan and Los Piños Rivers in northern New Mexico, the dam backs water up these stretches and into southern Colorado along the Piedras River. Along the San Juan and these smaller tributaries, a cluster of small Hispano communities had farmed this region along the border since the 1870s ("Navajo Unit").[5] With the CRSPA's authorization, the two hundred or so families living in the agricultural-based towns of Rosa, Los Arboles, Los Martinez, and Coraque were forced to abandon their homes, lands, and places of worship (Quintana, *Los Primeros* 1).

The disenfranchisement of those living in these communities is so compelling because while Glen Canyon Dam prompted a massive campaign from the Sierra Club and other sympathizers and has led to the

Rosa processional. *Before Obliteration by the Navajo Reservoir: Celebration of Santa Rosa Day, July 24, 1960*. Box 2, Folder 31, Frances Léon Quintana Papers, Center for Southwest Research and Special Collections, University of New Mexico Libraries.

proliferation of anti-dam discourse from this region lamenting the loss of wilderness and wilderness-related recreation, comparatively little has been said about the forced relocations occurring at Navajo Dam. Of course, the irony of arguing on behalf of wilderness in this region relies on the erasure of those living on the land. Addressing the failure to acknowledge the role that the US-Mexican War has played in the United States' wilderness ethic, ecocritic Priscilla Ybarra writes, "For the idea of wilderness to work, the U.S. had to erase not only the American Indian but also the Mexican American transformation of the natural environment in California, Nevada, Utah, New Mexico, Texas, Arizona, and parts of Wyoming, Colorado, Utah, and Kansas—all former territories of Mexico" ("Erasure" 136). To contrast these different views of nature, anthropologist

Rosa cemetery. *Before Obliteration by the Navajo Reservoir: Celebration of Santa Rosa Day, July 24, 1960*. Box 2, Folder 31, Frances Léon Quintana Papers, Center for Southwest Research and Special Collections, University of New Mexico Libraries.

Devon Peña explains that "Mexican Americans tend to define nature as homeland, not natural resources and wilderness" (*Mexican Americans* xxxii). Foregrounding Glen Canyon's disappearance under the waters of Lake Powell as America's most tragic environmental loss instead of the history of Navajo Dam and the impacts upon the dispossessed further reinforces America's fantasy of a prelapsarian West.

While one anticipates such an oversight from the general public, it is surprising, but perhaps not unexpected, that the Bureau of Reclamation's history of the Navajo Unit written in 1998 largely skims over the impact of the dam and reservoir on the Hispano communities. In this document the author notes "two major construction obstacles" that included portions of the Denver and Rio Grande Western Railroad that passed through the area and the cemeteries that were discovered during the dam's excavation (Linenberger 12–13). Of the latter, the author inserts her discussion of the discovery and reinternments of the bodies within the larger discussion of government contracts, implying that the effort and money required to move the bodies took priority in the reconstruction of the dam's history over the actual account of those living in the area. And despite the fact that 288 bodies were exhumed and reburied from cemeteries in Rosa and Rio de los Piños, there is no acknowledgment that people actually inhabited the area at the time of the dam's construction and were relocated by government mandate—a point challenged by the extant photos of the towns before their inundation (13). The bureau's history presents the information about the cemeteries as though one had found the ancient ruins of a vanished people rather than the sacred space of recently vibrant communities.

While this report glosses over the history of the Hispano communities, a display at the Navajo Visitor Center located in Navajo State Park, created to provide recreational amenities to the dam and reservoir's storage and hydroelectric functions, briefly accounts for the displacement. In a series of illustrated panels tracing the area's history, the park addresses the Indigenous presence, Domínguez and Escalante's expedition, and the Hispano arrival in the San Juan Basin. In a panel titled *Rails to Riches*, viewers learn of the Denver and Rio Grande Railroad's run to profit on southern Colorado's mining boom and the subsequent creation of Arboles, a stop along the rail line at the junction of the Piedra River. The panel also notes the railroad's impact on the local economy in Arboles and Rosa. As for the dam's impacts on these towns, the panel offers a small photo of the reservoir with an arrow marking where Arboles used to be and provides a brief note about the flooding: "Arboles and Rosa were covered with water when Navajo Lake was filled. The original town of Arboles lies just under the surface a few hundred yards off of Arboles Point" (*Rails to Riches*). This small exhibit provides useful information about the dam's impact on local communities to those visiting the state park. Yet the signage and accompanying text offer little of the human experience—the actual story of what transpired and how those forced to move reacted to the monumental changes coming their way. Rather, it portrays a rather matter-of-fact, emotionless statement of what occurred. To piece together a more complete narrative of this displacement, I turn to a variety of texts representing multiple genres to better account for the conditions and responses surrounding this little-known event.

The story of Hispano displacement is fragmented, with a variety of materials offering small glimpses into the emotional toll the dam's construction placed on the residents of Rosa, Arboles, and the other communities flooded by the reservoir. Key pieces to the narrative are Frances Quintana's (Swadesh) *Los Primeros Pobladores* (1974), an ethnology of the

Hispanic Americans within the Ute homeland, the scant archival material upon which Quintana created that book; Charles Wilkinson's 1996 environmental history of the Colorado Plateau; and Caroline Arlen's brief article from 2003 in *High Country News*.[6] Quintana's work provides the foundation to understanding this event, as it offers the most extensive information about the communities' displacement and background for Wilkinson's chapter, which includes a few statements from Quintana's husband and his daughter, who experienced the flooding of Rosa firsthand. These interviews lead Wilkinson to conclude, "Many different and sharp edges of conquest lay beneath the impounded water" (*Fire on the Plateau* 187).[7] The conquest Wilkinson observes is most evident in Quintana's book and private papers. Her study of northern New Mexico's Hispano communities dedicates a few pages to the Navajo Dam affair, calling it a "death-blow" to those living in the San Juan River valley (*Los Primeros Pobladores* 126). Relocating to newly created towns like New Arboles in Colorado and other cities throughout the Southwest, the displaced reacted to this event in the interviews Quintana conducted "with the melancholy of untimely death" (*Fire* 195; *Los Primeros Pobladores* 126, 128). The results of this depression were far reaching and extended beyond an immediate sense of loss. Quintana explains that premature death, alcoholism, and increased rates of mental and physical illness maligned those forced to move, leading the author to compare their plight with the trauma of "the Utes during the early reservation days" (*Los Primeros Pobladores* 128).

While Quintana's text provides an important overview of this event, it lacks many details regarding the dispossessed's reactions, a fact that reinforces Nixon's claim about how India's "self-assertive modernity depends on rendering invisible stories of national exclusion" (158). However, her reflections in journals and the few examples from the towns' inhabitants that do exist—part of the work she calls "salvage ethnohistory," akin to the salvage archaeology conducted at places like Glen Canyon before the water flooded sites of historical significance—are useful in piecing together their reactions ("Western History Association Address"). In her papers, she notes the efforts of the local inhabitants to challenge the Bureau of Reclamation. She describes one of Rosa's residents who "proudly brought over the answers received from the Rosa landowners [*sic*] petitions. Sen. Anderson and Rep. Morris said they had referred the matter to the Bureau of Land Reclamation with strong recommendations that the method of appraisal be thoroughly reviewed and that a practice of generous and equitable appraisal be followed. They said nothing about the church" (Interview). Unfortunately for the people of these communities, the bureau and other government agencies did not honor these requests. According to Quintana, landowners received only $125 per acre for irrigated land that would typically earn up to $600 (*Los Primeros Pobladores* 127). In a 1961 radio interview, Quintana explains that many of the dispossessed are elderly and that because these are people who "want to live out where there is space, out where a large family group can function," losing their homes "is a terrible tragedy in their lives" ("Discussion with Members of the Tri-Ethnic Program"). José "Ed" Marquez corroborates the poor compensation they received, explaining in the *High Country News* article that "[the Bureau of Reclamation representatives] said they were going to buy our land.... If you refused to sell, they would just condemn your place and tear it down anyway. They bought you out, gave you so much time, and you had to pack up your little rags and leave" (Arlen).

Some of the possible reasons for the scarcity of documents describing the Hispano resistance to the dam building is the fear residents may have had of appearing un-American during the Cold War. Quintana suggests in the radio interview that these people had "their backs to the wall" ("Discussion with Members of the Tri-Ethnic Program"). Because of their poverty and inability to muster significant political support, the residents of Rosa and the other flooded towns likely feared even greater prejudice by resisting the dams viewed by many as strategic pieces in providing for the nation's defense. For example, in 1957, the engineer-secretary of the Upper Colorado River Commission explained to a Utah congressman that "the storage of water in Navaho Reservoir is essential to serve the rapid development of the same industries in the Four Corners area. The Rocky Mountain West is recognized as the great dispersal area to which the Government of necessity must look for safety in a nuclear war. Wyoming, Utah, Colorado, and New Mexico embrace most of the uranium production in the United States" ("Comments of Ival V. Goslin"). Speaking not too many years after Joseph McCarthy's witch hunts, one of Rosa's residents confided in Quintana after she urged the man to do a radio interview to publicize the communities' plight. She recalls, "He was all for it but he was afraid that too much vigorous and collective noise would be interpreted as 'Communist agitation'" (Interview). Because the Bureau of

Reclamation occupied a privileged position within American political circles at this time, it is no wonder that this impoverished resident from a backwoods New Mexico village would hesitate to make such a public case against a reclamation project so many in the region felt was essential to the United States' effort to deter the spread of communism and ensure global peace.

The events surrounding Navajo Dam read much like Frank Waters's *People of the Valley* (1941) and John Nichols's *Milagro Beanfield War* (1974), which present fictional accounts of Hispano water woes caused by Anglo development.[8] Although these narratives take place within the Rio Grande watershed just east of the Colorado, they are linked ideologically and materially, as similar forms of conquest threaten both the real and the fictional Hispano communities and as water from the Colorado River is diverted into the Rio Grande through the San Juan–Chama Project, one of the participating projects of the Navajo Unit. Thus, while these novels imagine the environmental injustices northern New Mexico's Hispano communities have experienced, Navajo Dam manifests the very tangible and destructive conclusions expressed in fiction. As a terrain of struggle, the San Juan Valley and site of Navajo Dam and Lake represent, as Quintana observes, a "national sacrifice area," which, like other natural resource issues in New Mexico suggest, "sacrifice[s] the lives of those most intimately attached to the lands and waters of the state, Native Americans and Hispanos" ("Land, Water, and Pueblo"). Thus, this reading of the Navajo Unit portrays a far more complex picture of western reclamation than the realization of the garden or wilderness myths that typically characterize the narrative of Colorado River dam construction. While Glen Canyon Dam has long stood as the centerpiece of environmental debates regarding the Colorado River and US rivers in general, and although Glen Canyon may still continue to be, as Abbey once suggested, the heart of canyon country ("The Damnation of a Canyon" 130), there is much to learn about present water use from the events surrounding Navajo Dam. As Caroline Arlen concludes her report on the emergence of Los Arboles during the drought years of the early 2000s, "One can almost hear the lively voices...blowing through the valley" (Arlen). Today's water challenges invite us to listen to and engage narratives of this particular literature of the lost to more fully understand the consequences of river development and their impacts on some of the basin's most vulnerable communities.

Arboles ruins emerge from Navajo Lake. Photo by the author.

Locating the Goodlife and *Querencia*

In the decades following the construction of Navajo Dam, Latina/o voices in the watershed have gained strength, adding new chapters to a growing body of discourse about their claim to the Colorado. The stories these voices relate move from the sense of loss that permeates those describing the Navajo Unit's impacts on their culture and connection to the San Juan Basin to more powerful declarations of the values that have sustained these resilient communities over many generations. Despite the marginalization that has kept these communities on the periphery of Colorado River management, it has also reinforced a way of life articulated through a rhetoric of the goodlife that permeates these emerging narratives. In her work on Mexican American environmental literature, Priscilla Ybarra identifies what she calls "goodlife writing," which "embraces the values of simplicity, sustenance, dignity, and respect.... [T]hey function to preserve mutually healthy relations among individuals and communities. The values in goodlife writing implicitly integrate the natural environment as part of the community, and thus cultivate a life-sustaining ecology for humans" (*Writing the Goodlife* 4–5). These principles erase the divide between nature and culture that has traditionally defined western nature writing reflected in the robust body of

Glen Canyon narratives. This fusion of community and environment built upon a foundation eschewing individual gain and excessive material pursuits reflects the type of attitudes and behaviors needed to reassess relationships with the Colorado River. Likewise, the goodlife approach offers useful methods to reinvigorate environmental discourse and practice, suggesting that Latina/o knowledges of the land long predate the mainstream American environmental movement. As Ybarra affirms, "One of the most significant things that environmentalists can learn from the decolonial writings of Chicanas and Chicanos (among other peoples of color) is the fact that we never needed to become environmentalists in the first place, and we therefore have an array of strategies at our disposal for how to live well with Earth" (28). Communities in the American West can learn much from listening to the stories of these groups who have cultivated responsible relationships with the natural world for generations.

Closely associated with the goodlife is a generations-old term that more pointedly addresses the Latina/o-Hispano nexus between nature and culture to express one's intimate relationship to place. With its roots dating back five hundred years to Spain, *querencia* "describes the inextricable link between the land and its culture," where "the culture develops through generations of adaptation to the land, and it evolves in a way that sustains the land" (Fisher 484). For Estevan Arellano, the late New Mexican poet, historian, and advocate for the Upper Rio Grande region and its small Hispano communities, *querencia* is "that which gives us a sense of place, anchors us to the land, and makes us a unique people" (35). Essential to maintaining this recognition of one's unique identity forged by a deep connection to place is ensuring that future generations have the opportunity to benefit from this special relationship. This unique take on sense of place, coupled with goodlife rhetoric, underscores the value of the Latina/o community's discourse to sustainable water use within the basin.[9]

Examples of Colorado River goodlife writing and *querencia* appear in the digital presence of the Nuestro Río, or Our River, organization. Launched in 2011, the website and Facebook page promote the Latina/o voice in the Colorado's future development. The Facebook page explains that "Nuestro Río is a network of over 19,000 Latinos in the West that use our collective voice to educate our communities about the history of Latinos and the Colorado River and to tell our story as we advocate for a healthy river for generations to come" ("About"). Emphasizing the large percentage of Latina/o people in the basin and the organization's members who "treasure our long-standing historical and cultural connections with the Colorado River," the website invokes *querencia* to establish their commitment to the region. Their homepage invokes *querencia* as well as the goodlife as it explains, "One out of every three people living in the states around the Colorado River is Latino and we must make sure that our voices are heard in the debates over how to preserve this valuable resource for our children and grandchildren" (Nuestro Río). The strong presence the Latina/o community has in the region demands respect and a voice in deciding the future outcomes of the river. The Facebook page includes a short video introducing the viewer to the organization's mission through comments by its Regional Water Caucus members, who include state senators, house representatives, and leaders from city councils. Joe Salazar, a Colorado House representative, recognizes not only that the challenges facing the Colorado will impact his state but that such concerns are "a problem affecting all of the basin residents" ("June 29, 2015"). Nuestro Río's sense of *querencia* extends beyond this basin-wide approach to include a commitment to the future of Latina/o involvement in the watershed, visible in the Nuestro Río Youth Leaders program, which teaches young Latina/o and American Indians about their connections to the river and how to protect it. Another short video emphasizes the organization's attention to future water leaders, this time produced with American Rivers to highlight a group of twenty high school students who traveled on the Colorado and then to Washington, DC, to meet with policy makers and politicians ("Nuestro Rio, the Colorado River"). While the website and Facebook page list few updates since 2016 and 2018, respectively, these online forums nonetheless provide a record of a growing effort by the Latina/o community to sustain not only their homelands but also future generations so as to bring this tributary voice more directly into Colorado River governance.[10]

An even more direct articulation of the goodlife and *querencia* in the Colorado Basin is the fourteen-minute film *Leche y miel* (Milk and honey) presented by the nonprofits Hispanic Access Foundation and American Rivers.[11] A recipient of various awards, the short film opens with Jose Gonzales leaving his home before dawn to drive to the fields where he will spend his day driving a tractor. His voice soon chimes in, speaking in Spanish of the drug and alcohol addictions he overcame and the plea he made to God to change his life. Gonzales then references the Old

Testament story of Moses entering the Promised Land, the land "where milk and honey flow," after wandering in the wilderness for forty years. As Gonzales speaks these words, the shot transitions to the sun rising behind a mountain, and the words "Yuma, Arizona" appear. Here Gonzales's voice returns: "But in this moment," he declares, "we are already in that land, because this land has all of that." A series of pastoral images underscore the bounty and beauty of this place in the early-morning light. The fields thick with different varieties of lettuce are juxtaposed with low-angle and close-up shots of the workers gently handling the sprouts during planting. Soft instrumental music overlays the scene, and Gonzales notes, "All these plants are grown here thanks to the Colorado River." A final shot of an endless sea of sprinkler heads watering a freshly planted field closes out the introduction.

This opening scene's attention to the workers who make their living in the Yuma area and the crops upon which so many people throughout the nation depend (the vast majority of nearly all leafy greens come from the border region during the winter months) establishes the tone for the film and its celebration of this particular culture and its relationship to the Colorado River. At the heart of its numerous vignettes are goodlife values bound by an "ethic of sustenance... and an ethic of simplicity" (Ybarra, *Writing the Goodlife* 6). However, the film also serves as a warning, as it opens the next section explaining how vital the river is to the economies and communities that rely upon this highly threatened resource for "jobs and food"—that is, the river is essential to the dignity and sustenance of these people. Thus, as the film relates the story of these people, it also becomes a powerful rhetorical tool to show their meaningful connection to the river. Three and a half minutes into this film, it shifts to showcasing various organizations, families, and individuals who depend on or have been shaped by the Colorado River in one form or another.

The religious theme invoked in the introduction continues as viewers meet worshippers at the evangelical Iglesia Betania and Victor Venalonzo, its pastor. Venalonzo states that the Colorado River is "the life of Yuma. We live, we have jobs because of the Colorado River." And because of the drought and diminishing flows, Venalonzo explains he has rallied with other Christians to be more proactive in recognizing their duty to be caretakers of the river, to better fulfill the command in Genesis to be good stewards of the blessings given them. Noting Yuma's majority immigrant population, Venalonzo underscores the relationship these people have to the river—it is what makes it possible for them to earn a living. Here, as in other parts of the film, the goodlife principle of sustenance permeates the discourse. *Querencia* also infuses the narration. Toward the end of the film, we return to Venalonzo's voice as new initiates into the faith are baptized in the river at Yuma's Gateway Park. The slow-motion immersions blend with Venalonzo's voice-over that asks, "If we don't take care of our water, if we don't take care of our rivers, what inheritance are we leaving our grandchildren?"

This preoccupation with the health of the river for future generations speaks to these people's *querencia*. They recognize the river's role in the formation of their religious and cultural identity and feel a close connection and responsibility to protect the river. Later in the film viewers confront what happens when a people lack or forget their *querencia*. Reflecting on his family's past in a community south of Yuma, Louie Gradias tells how his grandmother used to travel upriver on a paddleboat. "Now there's no need for a paddleboat," he says. "You just need some high boots and you walk across." He even recalls the "river running like an ocean," but "now it's just a trickle." For him and previous generations, the river was a central part of life in the Yuma Valley, a once abundant resource whose very existence—and we can read this to mean the existence of those living in the area—is now threatened largely by a lack of awareness of the river's plight. Thus, Gradias's lament about the river's future and Venalonzo's pleas to his congregation to consider its responsibility to care for and protect the river become powerful appeals to place, belonging, and sustaining the special relationship these people feel with the river that has given them so much. Their observations recall Estevan Arellano's elaboration on *querencia* wherein he states that "we didn't inherit the land from our parents; we have it borrowed from our kids" (37).

The film returns to Gonzales, who speaks of the "very hard work" of laboring in the fields, but concedes that it is "beautiful work," as it "brings us satisfaction that we can work to grow food for other parts of the world."[12] Upbeat acoustic guitar music and footage of farmhands picking crops in the dark accompany his words. The next segment shows the family enjoying an evening meal together and Gonzales's wife, Olga, speaking of the "beautiful and dignified" work they perform on behalf of their children to give them a better life. With laughter and smiles in the background, Gonzales asks, "What else can we ask of life and of God? We have everything." The film concludes with images of Yuma's Latina/o community members in the places they work. These are then followed by a closing appeal to the viewer: "Help us sustain these connections to the Colorado River" (*Leche y miel*). This final refrain melds with the individual portraits of Yuma's farmworkers to create a new vision of what it means to rely upon the Colorado. The connections reflect the spiritual role the river plays in the lives of this community, the opportunity to do honest and meaningful work to provide for one's family and food for the masses, the identity forged by a lifetime rooted in a place, and the desire to preserve its history for future generations. As goodlife examples are distinguished by "the everyday, where we live, where we work, and often a crowded array of characters rather than a solitary soul on an isolated retreat," *Leche y miel*'s various narratives, which portray these communities' claims to the river, stand in stark relief to traditional wilderness rhetoric and its associated discourses of leisure evident in so much Colorado River imagination (Ybarra, *Writing* 27).

Mexico's Colorado River Delta

Appeals to the goodlife and *querencia* define not only the Latina/o communities north of the border but also the Mexican perspectives in the Colorado Delta, a place of great ironies, loss, and renewal. Charles Bergman, one of the most engaged authors with the region today, explains, "The delta has always challenged westerners.... It's the hottest place in North America.... It gets only a couple of inches of rain per year. The delta was, ironically, one of the first places in North America that Spaniards explored. But it has remained one of the least known" (17). Mexican scholar Jose Trava considers this irony, noting that river deltas are typically magnets for development and some of the first places settled. In the case of the Colorado, however, the delta has largely remained an enigma to the millions of people who rely on Colorado River water every day (172). Spanning an area one hundred miles long and eighty miles wide, the delta stretches southeast from California's San Gorgonio Pass just northwest of Palm Springs down to the river's terminus in the Gulf of California (Fradkin 333). It covers some of the most extreme landscapes in North America. However, in a great twist of irony created by the massive irrigation projects that have made the desert bloom, the region is the ultimate realization of the garden myth. It boasts some of the most fertile and productive agricultural valleys and irrigation districts in the United States and Mexico, including Imperial, Coachella, Palo Verde, Wellton-Mohawk, and Mexicali throughout southeastern California, southwestern Arizona, and northern Baja California and Sonora. While these efforts to reclaim the desert have made many areas in the upper delta very profitable, they have also substantially altered the river and its impact on the lower delta. With decades of dam building cutting off the river's annual flooding and even its exit to the sea, what once was a lush and vibrant ecosystem is now in many areas a desiccated expanse of sand flats and stunted vegetation.

When considering the delta's pre-dam years, many turn to Aldo Leopold's selection "Green Lagoons" from his classic on conservation, *A Sand County Almanac* (1949). Chronicling a 1922 hunting trip with this brother, Leopold identifies the delta's rich biological diversity as they float languidly along in their canoe, taking note of the egrets, cormorants, avocets, willets, mallards, widgeons, teal, bobcat, raccoons, coyotes, deer, and tracks of "el tigre," the elusive jaguar (151). Of the river itself, he writes, "On the map the delta was bisected by the river," he explains, "but in fact the river was nowhere and everywhere, for he could not decide which of a hundred green lagoons offered the most pleasant and least speedy path to the Gulf" (150). Amid this resplendent fecundity, Leopold perceives—like Jose Gonzales decades later, but in a very different context—a "milk-and-honey wilderness" (155). His observations suggest that he had half of the story correct. Indeed, the delta boasted one of the continent's more diverse ecosystems. Yet it was no untrammeled wilderness akin to what so many later generations of recreationists wanted to find in Glen Canyon; it was home to Indigenous peoples for millennia. Despite Leopold's failure to recognize the delta's Indigenous presence, his account offers an important glimpse regarding the ecological potential of the region and the impact large-scale agriculture has had on the delta.[13]

Inspired by such accounts as Leopold's, many texts over the past couple of decades have documented the region, providing a glimpse of what was and pointing fingers at who's to blame for the river's demise in this area.[14] No one culprit exists, although the US government and its seven basin states share the lion's share of the responsibility. As Bergman points out, in terms of who will ensure that adequate amounts of water once again appear in the delta, "The inevitable answer, the controversial answer, the incendiary answer is obvious: the United States" (19). However, the environmental challenges that the river faces today in the delta extend far beyond current management and use practices that have ensured that the Colorado dries up before it reaches the sea. A general attitude of ownership has prevailed north of the border, which has drastically compromised the river's integrity and the livelihoods of those in Mexico who depend on the river's health to survive. This untenable position also stands in stark contrast to present local perspectives on managing the region's natural resources that gesture toward greater collaboration between organizations.

Yet the problems that we face within the watershed extend far beyond what may be viewed as xenophobic laws and shortsighted regulations. Bergman astutely observes that the challenges hampering the river, particularly within the delta, are issues of imagination and language. In his conclusion to *Red Delta* (2002), one of the most comprehensive of the recent examinations of the region, he confides, "I have come to believe that this question of language—of what words we use for the delta, what metaphors we apply—is fundamental.... We have made the delta into an 'other,' an alien place, not seeing all the while that it was already home to people and creatures. Through our metaphors, we may have imposed ourselves upon the physical landscape, but we have not yet learned to live there" (279–80). As Bergman rightly observes, a crisis of representation mires the Colorado River delta region, as numerous texts represent the region as a wasteland and a place of little value. Bergman's repetition of "we" suggests that the trouble about representation derives from a US-centric perspective toward the delta, wherein it has been viewed as a blank slate upon which to project any number of denigrating or idealized representations. However, as Bergman's statement implies, other ways of understanding the delta must exist, for the delta has supported human communities for thousands of years. In order to learn to live in this particular region and the river that formed it, we must move beyond the debilitating perspectives that have "othered" the delta to offer more productive means of addressing this ecosystem's many challenges.

The othering of those in the delta is part of the United States' turbulent history with Mexico. In his examination of the delta region's political ecology from 1940 to 1975, Evan Ward observes, "A theme that unifies the history of the Colorado River Delta" is "the conquest and control of land and water" (4). Beginning with the Spanish arrival in the region in the 1500s as described in chapter 2, the delta experienced wave after wave of newcomers seeking to harness these resources. Following the United States' victory in the Mexican-American War, Mexico ceded millions of acres of its northern lands to the United States through the Treaty of

Guadalupe-Hidalgo in 1848 and the Gadsden Purchase in 1853. Well into the twentieth century, John O'Sullivan's rallying cry of Manifest Destiny, uttered on the eve of the United States' war with Mexico, continued to play out again and again as these two nations struggled to equitably apportion the Colorado.[15] When the seven basin states met to divide the Colorado in 1922, Mexico's water needs seemed to merit little more than an afterthought. The Colorado River Compact states, "If, as a matter of international comity, the United States of America shall hereafter recognize in the United States of Mexico any right to the use of any waters of the Colorado River System, such waters shall be supplied first from the waters which are surplus over and above the aggregate of the quantities specified in paragraphs (a) and (b); and if such surplus shall prove insufficient for this purpose, then, the burden of such deficiency shall be equally borne by the Upper Basin and the Lower Basin" ("Colorado Compact"). The important term throughout this entire section is the conditional "if" that precedes statements about sharing the water. The document's drafters left little question as to their position on Mexico's rights to the Colorado. If there is surplus water in the system, Mexico can have it. If not, then the Upper and Lower Basins will figure out how to deal with the shortage, ensuring that prior obligations between the Upper and Lower Basins are met. The conditional language also implies through its connection to the word "any" that Mexico has no right to the water unless the United States decides to grant its supposed neighbor such rights. In statements indicative of the general attitude toward Mexico's claims to the Colorado, Arizona congressman Carl Hayden once argued that Mexicans had "no right, legal or moral," to the river (qtd. in Hundley, *Dividing the Waters* 63).

Such attitudes prevailed in subsequent years with the passage of the Boulder Canyon Act in 1928, which paved the way for a host of projects that would forever alter the Colorado and the livelihoods of those along the lower stretches of the river. This act aimed "to provide for the construction of works for the protection and development of the Colorado River Basin, for the approval of the Colorado River compact, and for other purposes" ("Boulder Canyon Project Act"). With protection and development at the heart of the act and the states' desires, Mexico once again took a backseat in these discussions. While Section 4 apportions the water between Arizona, California, and Nevada, it binds Mexico's claim on the river (assuming, as few did during this time, that it had any rightful claim on the Colorado) within a number of stipulations about when and how such water deliveries should be made. In terms of the Gila River, the Colorado's last main tributary in the United States (once the border between the United States and Mexico before the 1853 Gadsden Purchase), the act states that "the Gila River and its tributaries... shall never be subject to any diminution whatever by any allowance of water which may be made by treaty or otherwise to the United States of Mexico." The act's penultimate section shows similar disregard: "Nothing in this Act shall be construed as a denial or recognition of any rights, if any, in Mexico to the use of the waters of the Colorado River system" ("Boulder Canyon Project Act"). What appears to be very much an afterthought, this reference to Mexico functions to allay initial fears about what this new infrastructure along the river would mean for downstream users in Mexico. At the same time, however, this provision emphasizes once again that Mexico may not even have any rights to the use of the river. Thus, in taking a highly ambiguous stance in regards to Mexico's water rights, the United States attempted to prolong its control on the river for the indefinite future.[16]

In 1944 the United States and Mexico finally took a significant step toward reconciliation when they signed a treaty regarding the Colorado River, Rio Grande, and Tijuana River. Facilitated through the newly formed International Boundary and Water Commission (IBWC), this treaty granted Mexico a 1.5 million acre-feet guarantee to be augmented by another 200,000 acre-feet in surplus years ("The Mexican Water Treaty").[17] While appeased by this guaranteed amount, Mexico soon discovered that water quantity was not their only concern. Because of the high salt content present throughout the watershed's alkaline soils, the Colorado has always carried a significant salt load on its way to the sea. But with increased agricultural operations, primarily in the Lower Basin, and the constant need to flush salts from the soil so as to not poison the crops, the river's salinity rapidly increased. One of the primary midcentury culprits, which added to Mexico's water woes, was Arizona's Wellton-Mohawk Irrigation District just east of Yuma. Comprising two fertile valleys along the Gila's southern border, these irrigation districts channeled huge amounts of agricultural wastewater back into the Colorado with salt levels almost three times the Colorado's average level (*Dividing the Waters* 173). What seemed like an easy way for the United States to meet its 1944 treaty obligations with Mexico proved disastrous for those farming downriver. Again, heated discussions between the nations emerged. Fortunately, for Mexico, there were those basin states in the United States sympathetic to this developing issue and that were instrumental in working toward a solution (177). Finally, in 1973, the International Boundary and Water Commission amended the 1944 treaty with Minute 242 that forged, as the title of the document suggests, a "permanent and definitive solution to the international problem of salinity in the Colorado River" ("Minute 242").

While the delta's plight has become a more widespread issue of concern throughout the watershed and beyond since the Colorado's high-water years in the 1980s and 1990s when the river last reached the Gulf of California on its own accord, the people living in the Mexican delta have long known the problems associated with this severely taxed river. Unlike many of their northern counterparts, however, their writings and representations of the region depict the delta in a much different light to provide the images Bergman stresses. Rather than perpetuate the paternalistic discourse evident in the United States' historical interactions with Mexico over the Colorado, or render invisible the delta's inhabitants through appeals to uninhabited wilderness, the Mexican voices initiate a variety of discourses that challenge these particular perspectives and approaches. In doing so, they represent the delta in ways that are both new and necessary to those rooted in so many Anglo-American portrayals of this region. The metaphors and images Bergman demands do not come from a Mexican version of Edward Abbey or any other environmental and nature writer who decries the delta's demise, however. While such alternative representations might be expected coming from an author, genre, or national tradition dedicated to using more poetic language to rethink humanity's place in the natural world, the Mexican voices locate their solutions within discourses and genres focused on action and tangible results.[18] As they invoke alternative metaphors and emphasize the goodlife and *querencia*, one finds the region cast in an alternative light that broadens our vision about what this place has meant to those who have inhabited the delta for generations.

AEURHYC and Delta Conservation

These alternative metaphors and ideas about the delta exist in a variety of place-based genres and discourses that challenge the long-standing perspectives—primarily emanating from the United States—that have marginalized this place and its inhabitants. Among these voices is the Asociación Ecológica de Usuarios del Río Hardy y Colorado, or the Hardy and Colorado Rivers Ecological Water Users Association. This organization represents a movement in the early 2000s to restore the delta to a vestige of its former splendor. Committed to the physical reconstruction and rehabilitation of significant portions of the delta, AEURHYC's founding documents demonstrate an equal preoccupation with reimaging the delta as a living, productive, inhabited place worthy of attention. Supported by other delta-based perspectives emanating from government, academic, and nonprofit representatives from the United States and Mexico, AEURHYC's 2004 "Plan Estratégico," or Strategic Plan, is largely a procedural document emblematic of the instrumental discourse one expects from a document outlining an organization's purpose and goals.

However, the plan's heavy emphasis on poetic discourse evident in its preface and repeated appeals to the goodlife, *querencia*, and collaboration galvanizes support and justifies the organization's role as a necessary and valued player in the broader dynamics of Colorado River politics. Its use of narrative ultimately functions as a powerful rhetorically motivated discourse to persuade readers to view the region as a place worth preserving, as do the perspectives articulated by Nuestro Río, those featured in *Leche y miel*, and the opponents to Navajo Dam's construction. As Beale reminds us, the purpose of rhetorical discourse is "to influence the understanding and conduct of human affairs" as it "operates typically in matters of action that involve the well-being and destiny of communities" and "in matters of value and understanding which involve the communal or competing values of communities" (94). It is within this confluence of rhetorical, instrumental, and poetic aims that we discover the range of approaches and mediums used to advocate for the delta, its people, and biotic communities. This breadth of rhetorical activity casts the long and often antagonistic relationship between the United States and Mexico over water use as a powerful environmental justice issue that is shifting the tides of how future binational water issues may find redress.

In response to the need to protect the rivers and wetlands throughout the delta, concerned citizens of the region began to organize. Formed in 1999 after seven years of collaboration with Pronatura, Mexico's foremost environmental organization, AEURHYC represents a diverse body of citizens who advocate for the delta's preservation and lead initiatives to involve others in the region's fight for existence (Hinojosa-Huerta 20). Though small, this group developed its initial "Plan Estratégico" for 2001 and later revised it for 2004–6 to outline how to restore the Colorado River, the Hardy River (a small tributary to the Colorado in this region), and the surrounding delta ecosystem. Though brief, the 2004 plan powerfully shifts representations of the delta as a wilderness/wasteland to a living, productive, inhabited place.

The 2004 revision showcases the organization's significant development and push for local restoration efforts over the previous three years. The 2004 plan's cover page depicts the name of the organization above a photograph of the Río Hardy at Campo Ramona, an ecocamp that often hosts American hunting and fishing enthusiasts, as it looked in May 2004. This photo depicts a wide stretch of the river surrounded by thick foliage on both sides, with mountains disappearing in the distance and a dock emerging from the photo's left. Superimposed on the photo are the words that state, when read in English, "working for the conservation of the Colorado River Delta" (AEURHYC 1).[19] Unlike Leopold's vision of the delta, this photograph blends the beauty of a well-watered locale with the actual human presence of those who live in the region, evident by the dock at the water's edge and the portion of someone's arm that appears in the lower left-hand corner. The human presence and development along the river reinforce the idea of work that the caption on the photo suggests. At the same time, however, their presence balances with the view of the river and mountains that recede into the distance. Humans neither overshadow nature indicative of a prodevelopment ethos, nor do they shrink before it as so many Romantics do in their pursuit of the sublime. Cropped to foreground the river while also allowing the elements of civilization to appear, the photo embodies AEURHYC's vision for what the delta can become: a place where a reliable presence of water can bring back Leopold's "green lagoons" and promote sustainable economic and environmental practices.

Whereas recreational and aesthetic interests promote the preservation of wilderness ideals in many of the basin's federally protected areas, the delta is a place where needs beyond these ideals are put in conversation and where a diverse set of stakeholders comes together to discuss how these needs will play out in the region. As articulated in the plan, AEURHYC's mission is "to conserve and restore the ecosystems of the Colorado River Delta and promote the wise use of its resources for the realization and continuation of activities that will allow for the economic, social, and cultural development of the local communities" (7).[20] This notion of "work"—like that expressed by Gonzales in Yuma—becomes a key term throughout AEURHYC's 2004 plan, as its members have their home in the delta; they are working primarily to preserve their livelihood and are concerned about how environmental restoration can prove economically sustainable for local communities and business enterprises. Work is further reinforced by a photograph on page 9, which depicts two individuals loading a wheelbarrow. The caption explains their restoration efforts at Campo Muñoz, another one of the delta's tourist camps.

The plan's most unique feature is the preface by Germán Muñoz López, a founding member of the group, which expresses the organization's desire to advocate for the plight of the region and its people. Rather than launch into the organization's goals, which would be more characteristic of instrumental discourse, this introduction adopts a reflective narrative of the river and delta, and thereby relies on pathos to underscore the ecological devastation that has shaped the area and the hope that can come through communal efforts. Thus, by relying on poetic discourse that emphasizes reflection and contemplation and its appeals to the reader's sympathies, López's introduction also develops the document's rhetorical aim in calling for changing and underscoring the association's values. López begins by writing, "There are few of us left who have seen the Colorado River in its splendor before its flow almost completely stopped. Most of us, now members of AEURHYC, had our first encounter with the Colorado River during the 1950s, while others like members of the Cucupá tribe have been here for hundreds of years."[21] This opening declaration establishes the human presence within the area and the relationship that the Mexican farmers and Cucupá Indians have had with the river for generations. It also respectfully acknowledges an Indigenous presence here, a move that the early water treaties along the Colorado failed to do. As the Mexican delta has long been viewed as a wasteland in the post-dam days, López's opening remarks locate a strong human connection to this region. From this point he continues to describe the river before the dams and heightened demand greatly reduced its flow, highlighting the river's strong and turbid currents that it was so well known for by early visitors to the region (2).

With the river's historic characteristics in the forefront of the reader's mind, López again returns to those who call this region home, explaining, "Since then we have been witnesses to how the river has deteriorated and how this has affected our lives" (2).[22] By emphasizing the intimate connection to the river and their ongoing observations about its drastic declines and the impacts this has had on the livelihood of those living in the delta, López endows AEURHYC with a unique ethos, one based on firsthand knowledge of how this region has changed over time. This act of witnessing is a powerful rhetorical strategy that many environmental writers have adopted. Evident in works such as *Testimony: Writers of the West Speak on Behalf of Utah Wilderness*, which emerged in light of the state's intense debates about wilderness preservation in the midnineties, this tool has inspired other collections about wilderness areas in Alaska and Nevada that witness to the state's natural beauty.[23] At the heart of this act of witnessing, or "speak[ing] from the truth of our lives," as *Testimony*'s editors explain, is story (Trimble and Williams 3). According to rhetorician Sharon Crowley, who looks to Aristotle's emphasis on "exemplary narrative…that makes a point by illustration or comparison" to construct compelling arguments, story is "the most efficient means of garnering attention" (197–98). Because story often provides a more subtle approach to transmitting beliefs and values, it is no surprise that AEURHYC opens this plan with Lopez's testimony regarding the river and what he and others experienced as the river and delta have drastically transformed from places of plenty to entities in grave need of rehabilitation. While an audience may disagree with his assessment, ultimately his story is his own; he is the sole authority on his childhood recollections about the river. As such an authority, Lopez's words command attention and respect. At the same time, this witnessing emphasizes the perspective of one intimately familiar with the land, not just an interloper's opinion. Thus, the plan offers a unique insider viewpoint—a bioregional ethic of *querencia*—of what this river means to those living and working day to day in this place that addresses what was lost *and* attests to what the future can hold.

One of the most obvious ways in which AEURHYC appeals to *querencia* is the preface's consideration of future generations. López affirms AEURHYC's commitment to restoring and protecting the region so that they may "enjoy it as we once did" (2).[24] The repetition of this idea and the ongoing focus on this goal clarifies AEURHYC's dedication to place and the viability of those who will inhabit it in years to come. By appealing to *querencia* and the attention it gives to later generations, López's preface moves beyond the nostalgia and loss that he shares with Leopold to look instead to future possibilities where the river and delta can again flourish. López builds on AEURHYC's earlier successes to foretell a new, more promising era for the delta. Despite the years of decline, which could easily leave López and others hopeless about the area's future, they embrace optimism and the slow progress already made to improve conditions.

Such embedded appeals to the shared responsibility to future generations reinforce the intimately expressed connections the people have with the delta and with future members of the community.

With the intent of restoring the delta so that future generations may benefit from it as López and others before him have, López returns once again to personal narrative to underscore his authoritative position as one who knew the delta before it drastically changed. He begins, "I remember when in 1952 a sudden summer downpour flooded everything."[25] Shifting the focus from the collective "we" to "I," López shares personal recollections of the river that continue to hold significant meaning for him. He recalls how as a five-year-old he returned to the river after the floodwaters of the previous year had receded and marveled at the nearly twenty-foot-high riverbanks that had been carved out by the river's force. He then goes on to list the abundance of the delta's flora and fauna that once thrived in the riparian and wetland habitats along the river. Up until 1960, López recalls that the river was in constant flux as significant runoff continued to augment the river's load, making those areas bordering the river prime habitat for animals like beavers, lynx, foxes, raccoons, and the "gato grande" (jaguar), not to mention the domestic animals that roamed the area. Drawing on this trope of abundance as Leopold does to emphasize the historic and quasi-idyllic characteristics of the region, López briefly shifts the narrative to the dramatic changes that followed the benchmark year of 1960 when he notes that this was the first time that the delta's infamous tidal bore stopped its charge upriver. Like the pulsating tides that wreaked havoc on ships trying to navigate the Colorado's mouth into the Gulf of California, López's reflections oscillate between remembrances and critique. While hinting at the disappearance of these infamous tides, he describes the special relationships that those living in the delta had with the river. In a place where summer temperatures regularly exceed 100 degrees Fahrenheit, "the river's waters were our only respite during the long summer days."[26] He then relates the fun he had with others after a day's work when the people would go to the river for swimming contests and mud fights (2).

Shifting again to nostalgic recollections that underscore the vanishing of this place as he once knew it, López's work is like the tide that ebbs and flows to leave different impressions on the land. Where his use of nostalgia and a vanishing landscape work to set up his appeals to address future generations, they also allow him to unite the experiences of a people living in a remote section of Mexico with a broader audience. By sharing memories from his past, he invites readers to consider their own special childhood memories and connections to places of significance. At the same time, López's stories help the reader see how those in the delta are not so different from people in other places; they find great satisfaction in their home place and have an appreciation of the beauty that surrounds them. Inherent in all these appeals is the human presence. They do not celebrate a humanless past that is more apparent in Leopold's treatment of the delta. Instead, López celebrates both the native and the nonnative species that made their home in the region and implicitly acknowledges the human component through the reference to the domesticated livestock. And unlike Leopold who visited the region for sport, López is here because this is his *querencia*, his home place. Rather than a place to escape to, the delta is the land of his ancestors and descendants, a place to live, work, and experience the goodlife. Although he laments a lost past, López also looks to the future and what positive changes can come to the region.

However, López interrupts these halcyon memories to shift his attention to the main culprit in the river's demise: the United States. Having just outlined the bounty and beauty that once permeated this region, López shifts his focus to the effects of the United States' reclamation boom, which placed countless dams, dikes, and diversions in the way of a free-flowing river. According to López, with the river pooling behind Glen Canyon Dam from the 1960s to the early 1980s, the river could no longer counter the effects of the Gulf's tides and winds that formed giant sandbars across the river channel. When the United States had to release massive amounts of water in the early 1980s because of high precipitation rates in the Upper Basin, the delta region and the greater Mexicali Valley repeatedly flooded because of the blockade, which prevented the water from reaching its historic outlet. Damaging livelihoods and placing the greater Mexicali Valley in danger, the United States' manipulation of the river had drastic effects on those in Mexico (2–3).[27] Whereas the Colorado River literary tradition has largely neglected the injustice of these effects, López directly addresses them.

The numerous changes in the river and the delta over López's lifetime lead him to lament the region's economic and environmental losses. He writes, "It is very sad to see what was once a river with a very large flow be converted into a dirty drain that in places is only five meters wide."[28] From the nostalgic and celebratory descriptions of the river in López's opening paragraphs, the preface shifts to a more pessimistic view on the present state of the river noted through the blame placed on the United States and those "*criminales*" that had recently set fire to the vegetation along the Hardy River (3). Ecocritic Scott Slovic identifies how negotiations of praise and blame operate as powerful rhetorical strategies to alert audiences to environmental issues. From his analysis of American nature writing, Slovic concludes that authors employ "discrete rhetoric" or "embedded rhetoric" as they shift between "rhapsodic" (praise) and "jeremiadic" (blame/warning) registers to explore the natural world and provoke an audience to action ("Epistemology" 82–86). In López's case, the text relies on discrete rhetoric or a "relatively clean split... into rhapsodic and jeremiadic halves" rather than the more integrated embedded rhetorical approach to provide a view of the delta and argue for its preservation (88).[29] While an embedded approach may be more effective in other situations, López's preface ensures that the audience understands the drastic changes that have characterized the river over the past few decades as a result of the United States' involvement and control of the river.

To further underscore the need to work to restore the river to a healthier version of its former self, López concludes the preface reasserting the connection the local people have with the river and delta, their role as witnesses to the delta's changes, and the charge to other groups and society as a whole to unite with AEURHYC. Shifting again his pronoun use, this time from "I" back to the collective "we," López states that it is upon the river that all those living near it "survive." He expresses hope that things can change with the involvement of others, but insists that such participation is "urgent." With such help, López believes that the area can be rehabilitated to the point that the wealth of resources that once characterized it and of which he and those in the region are "*testigos*," or "witnesses," can return once again (3).

I consider this preface at length because it occupies such an important place within the overall plan. As this is the first piece of information a reader encounters after the title page, it shifts one's expectation away from reading a more instrumentally motivated text that outlines procedural information about organizational structures and fiscal reports to one that centers on the specific values and desires of this particular community. Although instrumentally motivated discourses outline "the governance, guidance, control, or execution of human activities," or, more simply, the actions a group should perform, such discourse lacks the persuasive power that rhetorically aimed discourse carries that can galvanize support and the desire to act (Beale 94). To articulate AEURHYC's role in the delta's restoration, López relies on a balance of celebration, nostalgia, warning, and invitation and the repetition of images and phrases that evoke this group's *querencia*. Even though this use of epideictic rhetoric that includes praise and blame suggests Beale's poetic realm of discourse, López uses it to argue for the viability of this organization, their needs in protecting their lands and livelihoods, and, ultimately, a new reality: a greater Mexican presence in Colorado River governance.

Apparent in the sentiments expressed in this chapter's epigraph, water and land are resources that deserve protection and preservation. Water is not a commodity to be sold, as the Spanish *dicho* suggests, and neither is land something to be developed to the point that open space becomes only a memory. But in the case of the delta, perspectives that only foreground a wilderness appeal are ultimately detrimental to AEURHYC's cause. For too long the delta has been a blank spot on the map. While such "invisibility" is highly attractive to those seeking a refuge from more civilized regions, lack of awareness about the close link between the land and its inhabitants has compromised the freedom of those actually living in the delta. While a return to those conditions that Leopold praised—the endless maze of water and vegetation—would certainly benefit those living in the area who depend on a reliable presence of water, AEURHYC's position represents a clear environmental justice concern where the needs of people directly linked to a specific locale need to be balanced with those of the land.

AEURHYC's efforts have played an instrumental role in bringing much-needed exposure to this region and in establishing a governance model based on collaboration and cooperation among multiple constituencies. However, this community organization represents just one voice—albeit a very unique one—emanating from the delta today. In the spirit of what Craig Childs experienced as he entered the desert for answers to understand desert water, López, delta researchers, and the region's advocates have viewed the river and its surrounding land as the primary catalyst for creating the binational approach that currently governs many of the area's restoration and management efforts. Francisco Zamora-Arroyo, director of the Sonoran Institute's Upper Gulf of California Legacy Program, and a group of Mexican and American advocates and scientists trace the shifting attitudes by the United States and Mexico to the high-water years on the Colorado during the 1980s and 1990s. As a result of the flooding that occurred in the delta during this period after years of little to no water reaching the Gulf, the delta responded, prompting numerous individuals on both sides of the border to "explore the possibility that it could be restored" ("Collaboration" 871). Through the delta's natural regeneration during this period, many individuals expressed hope that in fact something could be done to bring the river and delta back to a healthier version of their former selves. AEURHYC represents one of the most prominent Mexican voices bringing about this transformation.

However, as this binational group of authors concludes, "Water alone is not enough to restore the Colorado River Delta" (884). Despite the periodic flows that do make it to the sea and provide necessary moisture for the various wetlands throughout the area, much more has to be done to ensure that the delta ecosystem that presently exists not only survives but thrives into the future. Clearly needed is an ethic that values the land and a people's rightful connection to that land. Again, Estevan Arellano makes a salient point about this relationship: "If we want the land to be taken care of properly, . . . [w]e must have people living on and from the land who are able and willing to take care of it. . . . We have to understand that we cannot save the land and water apart from the people or the people from the land and water. To save either, we must save both" (35–36). While Arellano addresses his comments to the situation in the Rio Grande watershed, his observations aptly describe the delta's situation as well as water crises everywhere. The Lower Colorado River and delta's survival are inextricably linked to how humans in and out of the region decide to view these natural entities. And as has been discussed at length, the fate of the people living in the region is clearly contingent on restoring and preserving the delta's environment. Fortunately, in the spirit of AEURHYC's motions to bring disparate groups together, a host of voices on both sides of the border advocate collaborative and cooperative measures between the United States and Mexico to define how the land and river, and the people and economies dependent on them, can prosper well into the future.

Since the 2004 plan, the two nations have taken major strides toward restoring the delta's ecosystems and human communities. In November 2012, the IBWC signed Minute 319. Among the various stipulations of this agreement was included the historic and experimental "pulse flow" of 105,392 acre-feet of water the United States would release in 2014 to mimic spring floods in an attempt to rejuvenate riparian habitat within the delta ("Minute 319" 14). Since the late 1990s, the river had not reached the sea, so when the Morelos Dam's floodgates opened in March 2014, delta stakeholders and river lovers on both sides of the border rejoiced.[30] The IBWC's final report submitted in 2018 regarding the effectiveness of these experimental flows identified varying results, but they pointed overall to the great potential regular water allocations could make to the delta ecosystem.[31] Building on the opportunities to improve the delta outlined in Minute 319, the two nations cemented their commitment to pursue restoration efforts in Minute 323. Signed in 2017, the document highlights "a shared desire to maintain existing binational work groups or to establish additional binational work groups to implement various elements of this Minute" ("Minute 323" 2). This mutual interest in collaborating for the benefit of the delta and for regulating water deliveries across the border reflects a growing recognition that the health of the river and its many communities must include cooperation and effort from many stakeholders. Set to expire in 2026, Minute 323 offers hope for the delta's future.

In the 1961 radio interview of Frances Quintana and one of Rosa's residents, the announcer exclaims—clearly moved by the loss the Hispanos were experiencing as Navajo Reservoir's water flooded their homes, businesses, and church—"people should know about it, know what is going on down there, and possibly something can be done in future cases like this—make a test case out of this" ("Discussion with Members"). Whether the "down there" refers to what occurred at the Navajo Dam site or south of the border in the Colorado Delta, the stories from these terrains of struggle bear repeating in order to bring awareness to the history of slow violence by Colorado River dams that have shaped Hispano, Latina/o, and Mexican communities throughout the basin. These histories help highlight the impact of decades of dam construction along the river, both to the river itself and to the communities living alongside it. Such histories also illuminate the ways in which wilderness narratives perpetuate the erasure of the Hispano, Latina/o, and Mexican voices in the watershed, particularly as such places like Glen Canyon continue to draw attention for the losses there. Finally, as these tributary voices tell their own stories of loss, belonging, hard work, and hope expressed in the goodlife and *querencia*, the discourses upon which we might draw today to ameliorate the attitudes and actions of the past emerge as voices from the deep. Largely invisible to those north of the border, these stories rise to inspire new generations to speak up and take their rightful place in Colorado River governance.

CHAPTER FIVE

Indigenizing Water Justice Along the Colorado

Rhetorical Sovereignty, Cosmovisions, and Systems of Responsibilities

Doom hangs heavy over the western story as far as Indians are concerned, but there is little expression of the reality of their situation as continuing participants in the region.

—ELIZABETH COOK-LYNN, "Assigning Meaning to the Lewis and Clark Adventure and the Western Story"

ON THE AFTERNOON OF SEPTEMBER 15, 1962, a large crowd gathered on a bluff overlooking the newly completed Navajo Dam. Dancers from the Navajo Nation entertained the regional and national attendees who convened for the dam's dedication. The ceremony began with the Navajo Tribal Band playing the national anthem and a group of local Boy Scouts raising the flag. The bishop of Gallup, New Mexico, offered the invocation, after which a number of celebratory speeches by various officials championed the dam's construction and its imminent benefits to the American Southwest. Among those addressing the crowd were Paul Jones, chairman of the Navajo Tribal Council; Floyd Dominy, chairman of the Bureau of Reclamation; and Stewart Udall, secretary of the interior (*Navajo Dam Dedication Program*). In his comments, Secretary Udall lauded Reclamation's numerous participating projects as part of the Colorado River Storage Project Act (CRSPA), noting that more money had been spent in 1962 than ever before ("Remarks of Secretary"). He then turned to the primary purpose of Navajo's construction and the subsequent Navajo Indian Irrigation Project (NIIP), a key project of the CRSPA: developing the Navajo Nation and its natural resources.

The CRSPA stipulated that NIIP would irrigate more than 110,000 acres on the Navajo reservation. Of the irrigation project, Secretary Udall explained that it would result in "employment opportunities, economic stability, and better standards of living for the Navajo people." Udall framed these benefits to the Navajo people within frontier and early-twentieth-century conservation rhetoric. Aware of the region's rapidly growing population, Udall told his audience, "Growth is a never-ending process which depends ultimately on the use we make of all our natural resources. Yesterday, this posed no problem because resources beyond imagination lay westward for the taking. Now, however, the pinch of onrushing national growth is being felt." Concerned that the San Juan's waters ran toward waste as it flowed uncontrolled toward its confluence with the Colorado in Glen Canyon, and viewing the region's resources as ripe for development, Udall reminded his listeners of the nation's duty to utilize the West's resources to their full potential. "Never again can we allow ourselves the foolish luxury of waste," he remarked. "Never again can we relax our efforts toward the optimum development of all the resources of our land, nor can we allow the process of resource conservation and development to falter as was done in recent years" ("Remarks of Secretary").

As if to further underscore the import of Udall's words for the Southwest's future, the following day, September 16, saw two more significant dedications that, in the words of one official, "open[ed] scenic Indian Country that was never before accessible" (Valentine). The first was the dedication of Navajo Route One, a key section of highway between southwest Colorado and Flagstaff, Arizona. Known today as Highway 160, this new stretch of paved road offered quicker access through the Four Corners region.[1] With this new access point in place, the Department of the Interior joined the Navajo Trail Association, the Navajo Tribal Council, and the Ute Mountain Ute Council in also dedicating a monument at the Four Corners "as a matter of historical interest" (Udall, "August 23, 1962"). Following these dedications, the Navajo Nation and Ute Mountain Ute Tribe hosted a barbecue for an anticipated five to six thousand attendees as the culminating event in a two-day celebration marking the beginning of a new era for the Southwest and its Indigenous communities (Valentine).

These historic events capture a significant moment in the development of the Southwest's resources and the progress of the region's Indigenous peoples toward greater self-determination. Yet the years leading up to these celebrations also reveal important insights about the many struggles the Colorado Basin's Indigenous nations and tribes have faced in asserting their legal rights to natural resources. Once again Navajo Dam, its construction, and the realization of its associated water infrastructure are central to this complicated story of Indigenous rights, resource development, and environmental justice. Whereas the previous chapter suggests how Navajo Dam's construction represents a powerful terrain of struggle for the Hispano communities of northern New Mexico, when understood in terms of its relationship to the Navajo Nation and other Indigenous communities receiving Navajo Lake's water to fulfill federal treaty obligations, the dam becomes a more complicated symbol of environmental justice. It is here that we see two historically marginalized groups' claims to place pitted against one another. At the same time, however, the government's inability to fulfill the entirety of the promised benefits made to the Navajo and other Colorado River Basin tribes through the CRSPA and other federal legislation regarding Colorado River water rights suggests that even promising gestures toward a more equitable water-governance system continue to fall short for the basin's Indigenous communities. A narrative of possession and dispossession regarding Indigenous homelands and home waters throughout the basin constitutes the primary focus of this chapter.

Using the events surrounding the Navajo Unit's approval, construction, and ongoing issues as a springboard to initiate an examination of Indigenous water rights in the Colorado River Basin, this chapter adopts the "first in time, first in right" premise that governs western water. I use this principle to bring Indigenous knowledge about the Colorado River Basin and the natural world more broadly out of the mainstream's obscurity to reposition these tributary voices at the foreground of the region's water cultures. This decolonization of Colorado River Basin water knowledge employs Dina Gilio-Whitaker's call for "indigenizing environmental justice" to examine a number of texts representing various tribal affiliations and genres to consider how their particular use of narrative engages the historic and ongoing environmental injustices they have faced and continue to negotiate in their fight to preserve their sacred lands, identity, and access to reliable, clean water (26). Such a decolonization questions the utility of Walter Beale's discursive framework, which itself is deeply rooted within the Western tradition, as a means of understanding how Indigenous narratives operate. In the case of the texts examined herein, they adopt the various discourses of Beale's model to enact their own "rhetorical sovereignty," what Scott Lyons defines as "the inherent right and ability of *peoples* to determine their own communicative needs and desires in this pursuit, to decide for themselves the goals, modes, styles, and languages of public discourse" (449). This rhetorical sovereignty "requires above all the presence of an Indian voice, speaking or writing in an ongoing context of colonization and setting at least some of the terms of debate" (462). Thus, as the Indigenous voices in this chapter employ scientific, instrumental, rhetorical, and poetic registers, they work within the dominant discursive modes constructing mainstream Colorado River knowledge while manipulating these discourses and their respective genres to prominently establish their presence within this arena.

The rhetorical sovereignty these Indigenous voices claim demonstrates not only their expertise in addressing current water issues but also their ability to exert their "Indigenous eco-agency" to challenge the status quo and offer their own solutions rooted within their respective experiences and traditions (Adamson and Monani 9). And narrative—the stories of these traditions and connections to land and water—is central to Indigenous sovereignty in terms of both the language used to address environmental concerns and the on-the-ground efforts to control resources. As Cree scholar Stephanie Fitzgerald argues in her assessment of Indigenous women's writing about land dispossession and environmental justice, "It is narrative that creates the representation of the dispossessed, challenges the hegemonic invisibility of Native land dispossession in all its forms, and disseminates potentially useful and strategic counternarratives" (15). In this vein, my treatment of these narratives spanning Leslie Marmon Silko's novel *Gardens in the Dunes* (1999), Ofelia Zepeda's and Natalie Diaz's poetry, the Ten Tribes Partnership Tribal Water Study (2018), and the Grand Canyon Trust's "The Voices of Grand Canyon" digital project (2020) sheds greater light on the essential cultural, spiritual, and economic relationships that the basin's nations and tribes have with the Colorado River. Through these counternarratives to the West's dominant water ideologies and cultures, and one's willingness to rhetorically listen to their messages, the basin's tribal nations draw attention to past and ongoing struggles to secure equitable water access while amplifying their resilience and self-determination.

American Indian Water Rights

When considered in light of Native water rights, the "first in time, first in right" principle is one of the more ironic aspects of western water use. When the US Supreme Court ruled in favor of tribal water rights to Montana's Milk River in the landmark 1908 *Winters v. United States* case, it stipulated a priority date based on the date of the federal reservation's creation, even if the treaty or other decree setting aside tribal lands did not include water rights. As many of the West's reservations were created in the late eighteenth century, long before the Colorado River and its tributaries were diverted, the tribes often hold the most senior rights to this water. Despite such senior rights, the tribes have historically lacked the means by which to perfect or develop those rights, whereas settler communities, often aided by state and federal dollars, have controlled much of the West's waterways and their use. The language of the 1922 Colorado Compact in regards to the basin's twenty-nine federally recognized tribes is very revealing about where these "first in time, first in right" groups stand. In Article VII the compact states, "Nothing in this compact shall be construed as affecting the obligations of the United States of America to Indian Tribes" ("Colorado River Compact"). While this brief article acknowledges tribal reserved rights and the federal government's responsibility to protect them, the compact offers no other guidance as to how those obligations will be fulfilled at either the federal or the state level. Even Mexico receives more consideration in the compact than the sovereign tribes living throughout the Colorado Basin. The lack of further elaboration within this article or the larger compact reflects an overall disregard for the tribes and their senior rights during this time.

Such oversight continued throughout much of the following decades when the seven basin states worked furiously to appropriate their allotted shares outlined in the 1922 compact. When the Lower Basin states convened in 1928 to divide their allocation of 7.5 maf and to approve the construction of Boulder Dam, no mention was given to the tribes in the area. Nor did the 1944 treaty with Mexico address how the tribes with rights to the Colorado within the United States might impact how it delivered water to its southern neighbor. Not until the Upper Basin's 1948 compact did the tribes and their legal claims to significant portions of Colorado River water receive more attention. In the exact language of the 1922 compact, this document acknowledges the federal government's duty to the tribes. Yet it also briefly notes in Article XIV, which outlines the division of San Juan River water between Colorado and New Mexico, the tribes' "preferential uses" that will not figure into the water cuts the two states will make during potential water shortages ("Upper Colorado River Basin Compact"). Then in 1956, Congress ratified the CRSPA, which represents an important turning point for tribes on the Colorado Plateau, and particularly the Navajo, who became the primary beneficiaries of the Navajo Unit. Nonetheless, the legislation, like so many other issues revolving around Indigenous rights, is mired in a complex history arising from multiple stakeholder needs, limited resources, a bevy of federal laws, and efforts to balance a rapidly developing Southwest.[2]

A key provision of the CRSPA legislation outlined opportunities for the Navajo Nation to develop its water rights to the San Juan River and thereby improve its economic standing through the proposed Navajo Unit, which included a dam and reservoir. The legislation also stipulated that the secretary of the interior would "give priority to the planning of the reports" for a number of participating projects, which included a future irrigation project connected to Navajo Dam and Lake. Framed within a "doctrine of self-determination" as the bureau's history of the dam reveals, the Navajo Nation's efforts to improve its economic outlook hinged on Congress's approval of a future irrigation project to benefit its people (Glaser, "Navajo Indian" 13). In a 1956 letter dated a few months before Congress's approval of the CRSPA, Paul Jones, chairman of the Navajo Tribal Council, reminds Utah senator Arthur Watkins that "this development will be of utmost value to our people and it has our wholehearted support." Throughout the letter, Jones emphasizes the Navajo Nation's plans to develop its industrial and agricultural sectors and underscores the importance that the Navajo Unit will make to the reservation as it helps lift the people out of poverty. For Jones, the Navajo Unit's "extensive" benefits, which also included needed flood control, would aid in the "rehabilitation of the Tribe." He concludes his appeal to the senator stating that the Navajo people "are extremely anxious for approval of" the CRSPA legislation ("February 8, 1956").[3]

To further promote San Juan River development, the Navajo Nation produced a brochure titled *What the Colorado River Storage Project Means to the Nation's Largest Indian Tribe: The Navajos* to assist the public in understanding the current conditions on the reservation and the value that a Navajo-specific irrigation project would bring. This two-sided document

The Navajos and the Colorado River Storage Project. MSS 146, Box 29, Folder 16. Courtesy of L. Tom Perry Special Collections, Harold B. Lee Library, Brigham Young University, Provo, Utah.

includes three photographs capturing the hardship of collecting water and the difference reliable water infrastructure would make in transforming the reservation's desert soils into productive cropland. Accompanying these images is Jones's statement to the reader titled "New Hope for My People." In more direct language than he used in writing to Watkins, Jones explains that the Navajo Unit "will help us become self-supporting by enabling us to put to use the water that will irrigate our land and provide water for other purposes." Jones builds his argument in favor of Navajo self-sufficiency by noting the Navajo Unit's role in "providing the food we need to keep us from starvation," while improving the nation's economic outlook that will "raise our standard of living." Key to Jones's statement is the appeal to Americans to enable the Navajo to be autonomous in their ability to care for themselves. The document repeats this sentiment throughout, evident in the closing caption that reads, "Give us opportunity to help ourselves." Other appeals, including those to fulfill the 1868 treaty obligations to provide the tribe with agricultural and educational opportunities, further underscore the pressing need for the Navajo Unit (*What the Colorado River*).

At the same time the Navajo Nation outlined its right to the San Juan's waters, the state of New Mexico also looked to the river to address its post–World War II boom. Communities along the Rio Grande corridor such as Albuquerque and the state's numerous military facilities had already fully tapped the Rio Grande and therefore looked to the San Juan River to supply their growing needs (Glaser, "San Juan–Chama" 8–9). New Mexico therefore promoted the transbasin San Juan–Chama Project by which to bring San Juan River water across the Continental Divide and into the Rio Grande watershed.[4] When Congress approved the CRSPA on April 11, 1956, it paved the way for the Navajo Unit and a potential Navajo irrigation project to move forward as well as the San Juan–Chama Project to bring much-needed water to the Rio Grande corridor. While the approval of the CRSPA legislation was a monumental step in moving forward the Upper Basin's desires to develop its resources on par with what was occurring in the Lower Basin, it nonetheless opened the door for years of conflict about the details of constructing and funding a Navajo-specific irrigation project and the San Juan–Chama transbasin diversion.

Even though Congress had given the initial approval for the CRSPA, it would not approve funding for NIIP and the San Juan–Chama Project until Congress could see that the construction of both participating projects was actually possible (10). The ensuing years saw many attempts by the Navajo Nation and the state of New Mexico to negotiate how much water each party required, which eventually culminated in the 1958 Navajo Irrigation–San Juan–Chama Diversion bill. The bill required numerous efforts by various parties to bring attention to the Navajo's need for the water. Drawing upon earlier arguments about the Navajo's requests, Representative Joseph Montoya argued before the House's Subcommittee on Irrigation and Reclamation in 1960 that NIIP "will not provide just a temporary alleviation of the economic problems with which the Indian people within this area are faced, but rather a permanent solution for the economic betterment of the Indian people" ("Representative Montoya" 2). He further emphasized that the project's economic boost to the reservation would "go a long way toward making the Navajo people self-sustaining and provide them with the self-respect engendered through productivity brought about through their own toil and work on their own farms" (3). Two years later, Paul Jones, along with other tribal members, traveled to Washington, DC, in 1962 to testify before Congress about the project's necessity (Glaser, "Navajo Irrigation" 14).

Despite the Navajo's federally reserved right to hundreds of thousands of acre-feet per year of the San Juan River, the nation understood that it would have to make concessions to receive the federal dollars to construct the irrigation infrastructure. Eventually, the Navajo Nation conceded a significant amount of water to ensure approval for NIIP and the San Juan–Chama Project. In the end, it agreed to a delivery of 508,000 af/year to water 110,630 acres, which would then enable a diversion of 110,000 af/year for the San Juan–Chama Project (Glaser, "Navajo Irrigation" 9). It also guaranteed that the Navajo would share their water during times of water curtailments (11). Of the Navajo's willingness to compromise even when they had the full force of the law behind them, Representative Montoya noted in 1962, "The Navajo Tribe has made this very generous and far-reaching concession because the development of these resources is necessary to the very survival of the Navajo people" ("Statement 1962" 3). Painting their concession as evidence for a future of Indigenous-settler harmony, Montoya enthusiastically recalled, "New Mexico recognized that the Navajo Indians deserved great consideration and merited some entitlement to these waters. In reciprocity, the Navajo Indians came into these councils armed with the maxim known to most lawyers and dispensers of justice that 'He who expects equity must be prepared to do equity.' So, with this kind of spirit pervading these important consultations, we became one family satisfied with the final resolution that we would work in behalf of a Navajo Indian Irrigation Project and the San Juan Chama Project through the approached allocation of waters from our State's entitlement" (15). With such a "spirit" of collaboration and sense of familial unity, the two projects could finally move forward to meet the needs of many different stakeholders throughout New Mexico.[5] However, Montoya's optimistic recollection of these events belies the challenges the Navajo Nation would face in subsequent years in its efforts to construct NIIP.

Despite the significant efforts to give both projects the green light, the full implementation of NIIP continued to face ongoing struggles to secure funding and begin construction, whereas the San Juan–Chama Project moved forward with far fewer obstacles. The Bureau of Reclamation's own history of NIIP acknowledges the project's checkered past, noting that its "implementation is often criticized for poor planning, little or sporadic funding, lack of support, and an arduously slow construction process" (Glaser, "Navajo Indian" 2). In Frances Quintana's study of the San Juan's Hispano communities, she acknowledges the marginalization the Navajo endured as they awaited the fulfillment of their federally reserved rights. Referring to what she observed in the 1970s, she laments, "To date, not a single Navajo has benefited from the impounding of these waters. The only visible benefits seem to go to the owners of power boats who come from several states to enjoy the lake, and to the operators of the marina and campgrounds" (*Los Primeros Pobladores* 128).

While construction on the San Juan–Chama began in 1964, with all major projects completed in 1971, progress on NIIP lagged. Glaser explains that by 1970, the San Juan–Chama Project was two-thirds completed, whereas NIIP was only 17 percent under way (Glaser, "Navajo Indian" 13). In 1970 Joseph Montoya, now a US senator, wrote to President Richard Nixon questioning the lack of funds for NIIP despite the fact that they had been approved nearly a decade earlier. In one brief passage he captures much of the US government's relationship with its Indigenous citizens: "I do not understand our Government's paralysis with respect to our Indian population" ("Letter to President Nixon"). Summarizing various opinions about NIIP's slow progress, Glaser concludes, "State and Federal agencies manipulated Indian sympathies and priority rights to gain funding for the benefit of non-Indian projects" ("Navajo Indian" 16).[6] With federal dollars and attention directed first to the San Juan–Chama Project, the Navajo Nation received its first NIIP irrigation in 1976, years after the entire project was slated for completion (24). Remarkably, nearly sixty years after President Kennedy approved NIIP, the project still has not been finished. As of 2020, only eight of the eleven ten-thousand-acre plots originally prescribed in the NIIP legislation are completed. In February 2020, Navajo Nation tribal president Jonathan Nez petitioned Congress for more funds to move NIIP closer to completion ("Nez Testifies").

A Path to "Indigenizing Environmental Justice"

This treatment of NIIP and the Navajo Nation's ongoing struggle to develop their federally reserved water rights in light of the previous chapter's consideration of environmental justice and the wilderness movement illustrate the pervasive marginalization and injustice that continue to impact the country's Indigenous groups. The history of Navajo Dam and Lake and their respective projects underscores how serving the needs of the dominant culture pits historically marginalized communities against one another. As the Navajo Nation received approval to improve the reservation's economic and social health through irrigation and large-scale agriculture, this project displaced not only a group of Hispano communities but its own people whose lands were transferred to NIIP (Glaser,

"Navajo Indian" 26). The Navajo Nation also had to relinquish some of its claims to its legally allotted water rights for the San Juan–Chama Project in order to gain enough support for its own. "First in right, first in time" sounds like a great idea if you are at the head of the line, but for the Colorado Basin's Indigenous peoples, this idea is far more illusory than it suggests.

Such marginalization furthers our understanding of environmental justice within the watershed, as recent scholarship in Indigenous rights suggests that environmental justice is much more than the equitable access to natural resources. The Environmental Protection Agency (EPA) defines environmental justice as "the fair treatment and meaningful involvement of all people regardless of race, color, national origin, or income, with respect to the development, implementation, and enforcement of environmental laws, regulations, and policies. This goal will be achieved when everyone enjoys the same degree of protection from environmental and health hazards, and equal access to the decision-making process to have a healthy environment in which to live, learn, and work" ("Environmental Justice"). For this federal agency, environmental justice assumes that all individuals have a voice at the table to ensure a safe, healthy setting in which to live. The editors of *The Environmental Justice Reader* (2002) also identify similar objectives in their definition: "The right of all people to share equally in the benefits bestowed by a healthy environment. We define the environment, in turn, as the places in which we live, work, play, and worship. Environmental justice initiatives specifically attempt to redress the disproportionate incidence of environmental contamination in communities of the poor and/or communities of color, to secure for those affected the right to live unthreatened by the risks posed by environmental degradation and contamination, and to afford equal access to natural resources that sustain life and culture" (Adamson, Evans, and Stein 4). However, their position also identifies race and class as significant factors for those most prone to experiencing environmental injustice as well as the importance of one's spiritual health in a just world. This broader articulation, with its focus on how marginalized groups disproportionally bear the brunt of unhealthy living conditions, gestures toward the manner in which Native scholars address environmental justice and its impacts on Indigenous communities.

For scholars such as Dina Gilio-Whitaker (Colville Confederated Tribes) and Kyle Powys Whyte (Potawatomi), environmental justice for

Indigenous peoples means equal access not only to the earth's material resources but also to its spiritual elements that comprise essential components of Indigenous identity. Their respective arguments for "indigenizing environmental justice," as Gilio-Whitaker contends, suggest a significant reorientation for how dominant culture envisions human-earth relations and posit needed directives for reimagining Colorado River water use in the twenty-first century (26). Gilio-Whitaker asserts that the traditional notions of environmental justice are rooted in colonizing systems, which have negatively shaped Indigenous cultures: "The underlying assumptions of environmental injustice as it is commonly understood and deployed are grounded in racial and economic terms and defined by norms of distributive justice within a capitalist framework. Indigenous peoples' pursuit of environmental justice requires the use of a different lens, one with a scope that can accommodate the full weight of the history of settler colonialism, on one hand, and embrace differences in the ways Indigenous peoples view land and nature, on the other" (12). Under this alternative vision of environmental justice, the equitable access to natural resources is only the beginning of a more just future for Indigenous peoples. Rather, an Indigenized environmental justice accounts for the basic ideologies that have compromised equitable access to those resources and embraces alternative epistemologies reflecting the unique spiritual, communal, and even economic connections tribes have with the earth.

Therefore, in the case of the EPA's statement on environmental justice, Gilio-Whitaker notes its shortcomings as a "universalizing, unidimensional approach that fails to account for different histories" (36). In sum, "For environmental justice to be responsive to the needs of Native peoples it must be indigenized—tailored to account for their very different histories, relationships to the land, and political relationships to the State" (147). On this latter point, Karen Jarratt-Snider and Marianne O. Nielsen's recent work on Indigenous environmental justice clarifies how popular articulations of environmental justice aiming for racial parity in the access to clean, healthy natural resources fail to account for the fact that "Native American tribes are governments, not ethnic minorities." This important distinction maintains a different relationship between tribal nations and the US government, one based on the "Trust Doctrine," which emphasizes the federal government's "obligation to protect the interests of Indian tribes" (9). One of this chapter's objectives is to foreground some of these different histories in light of tribal nations' unique legal position to raise awareness of their utility and relevance to present and future Colorado River governance.

Kyle Powys Whyte's work on environmental injustice informs Gilio-Whitaker's assessment and helps further elucidate the challenges Indigenous communities within the Colorado River Basin and elsewhere face when it comes to asserting their claims to their historic lands and waters. While Whyte's vision of an Indigenous environmental justice relies on principles of access previously articulated, his notion goes much deeper to the fundamental relationships humans have with other humans and the natural world that have been nurtured for countless generations. For Whyte, environmental "injustice also occurs when the social institutions of one society systematically erase certain socioecological contexts, or horizons, that are vital for members of another society to experience themselves in the world as having responsibilities to other humans, nonhumans and the environment. Injustice, here, involves one society robbing another society of its capacities to experience the world as a place of collective life that its members feel responsible for maintaining into the future" (12.4). The relationships Whyte traces between various human and nonhuman entities both in the present and for the future encapsulate his "systems of responsibilities," and it is the "interference with and erasure" of these connections that constitute environmental injustice (12.25). The subsequent analysis of Native fiction and poetry, the Ten Tribes' Water Study, and "The Voices of Grand Canyon" digital project highlights some of these "horizons" and their associated discourses through which tribes assert their systems of responsibility as river stewards.

Other scholars have also emphasized these particular relationships in light of Indigenous land possession. In summarizing Diné (Navajo) and Annishinaabe/Ojibwe perspectives on nature and their contrast to romanticized notions of the "ecological Indian," Stephanie Fitzgerald explains that because "nature is linked to everything else," nature is understood "not on sentimental feelings or affinities for what, in American English, can be amorphous concepts," but on a system of "relationships and stewardship" (11).[7] Joni Adamson and Salma Monani extend this understanding of Indigenous connections with nature by framing these approaches within what they identify as "Indigenous cosmovisions," perspectives that "articulate dynamic epistemologies that have been negotiated over long histories (sometimes thousands of years), and may present sound ethical and scientific reasoning for ecological protection" (9).[8] Thus, rather than framing the natural world as the antithesis of culture, a trope that has long bedeviled western literature, the Indigenous texts examined in this chapter forge connections between nature and culture where concern for the land and water reflects a concern for the self, the home, community, and all relationships that unite the human with the nonhuman world.[9]

These articulations of responsibility to the earth and its inhabitants run counter to private property and the American West's prior-appropriation doctrine. Although prior appropriation implies "beneficial" use—intended to prevent the "waste" of water flowing in a stream that could otherwise be used for a public good such as agricultural or domestic use—it is beholden to the prevailing beliefs about how water benefits society. While sensibilities have shifted in the United States to extend beneficial use to include ecosystem management and aesthetic purposes in certain areas, western water law is still a system rooted in ownership and individual rights that often pits users and uses against one another. In my readings of the texts that follow, I examine how the narratives broaden the possibilities of beneficial use through systems of responsibility, which Gilio-Whitaker suggests in her own summation of these connections are "relationships of reciprocity based on responsibility toward... life forms." Such relationships then, as they extend beyond the human world and emphasize the value of all life forms, become provocative examples of "native resistance" (13). Expressed and forged through narrative to establish the authors' rhetorical sovereignty, these responsibilities, relationships, and cosmovisions define how these "first in time, first in right" Indigenous communities today seek to develop their Colorado River allocations and use their unique legal position and discursive strategies to emerge as water leaders in an age of growing uncertainty and scarcity.

The Colorado in Indigenous Fiction and Poetry

Laguna author Leslie Marmon Silko dramatizes the Indigenous connection to the Colorado River in her novel *Gardens in the Dunes* (1999), wherein she depicts the historic traumas that have severed tribal connections to place and to their life-sustaining waters. Set at the turn of the twentieth century, *Gardens* follows the experiences of sisters and members of the fictitious Sand Lizard people: Indigo and Sister Salt. Raised among the ancestral gardens near the Colorado River, the two girls are separated from their mother when US soldiers break up a Ghost Dance attended by a number of the Southwest's tribes as "white people got uneasy when they saw numbers of Indians gathered in one place" (24). Like so many American Indian children at this time, the two girls are captured and shipped off to a boarding school in Riverside, California. Soon after her arrival, Indigo escapes and manages to meet newlyweds Edward and Hattie Palmer, who adopt Indigo and take her abroad, where she is exposed to the history and culture of Europe. Sister Salt, on the other hand, leaves the school to work in a laundry and then at the Parker Dam construction site, where she conceives a child and deals with the hardships of single motherhood and living on the margins of white society. Eventually, the girls reunite and even take in Hattie, whose tragic demise following her separation from Edward finds her alone, destitute, and the victim of a brutal sexual assault. Despite the many challenges the girls face, they arrive at the end of the novel in good spirits and in control of their lives.

Silko adopts a variety of genres within *Gardens* to address, as Yeonhaun Kang suggests, the "American West's past, present, and future surrounding the history of settler colonialism and its link to environmental and social injustice in the United States and beyond" (104).[10] Not only does Silko do this by casting independent Indigenous women as her protagonists, but her attention to water and its impact on gardenscapes in the US West, on the East Coast, and in Europe reveals how she critiques western American land-use practices at the turn of the century. This attention to water within the novel, John Purdy notes, speaks to Silko's efforts to "enlighten some fundamental differences between cultures, . . . as an investigation of European values as revealed by colonialism" (150). Silko's rhetorical sovereignty develops from the multiple critiques of gender, race, and nature throughout the novel, which reveal the systems of responsibility the sisters have to their desert homeland.

Early in the text, Silko develops the girls' connections to their home gardens and the systems of relationships with the various organisms that inhabit these places. Prior to the girls' abduction, Grandma Fleet instructs them about the importance of the desert gardens and the "precious moisture from runoff that nurtured the plants." The girls tend the plants "as if the plants were babies" and learn to act respectfully around them, as "the plants listen." Grandma teaches them to "always greet each plant respectfully. Don't argue or fight around them plants—hard feelings cause the plants to wither." Grandma's reverence for the gardens comes from her knowledge of the gardens' origin, which "had always been there," and she passes this knowledge on to her granddaughters (14). She tells them how Sand Lizard planted the gardens long ago and instructed the people how to share and cultivate the gardens so they would survive. Grandmother reminds the girls that the gardens "could be counted on for sanctuary" (14–15). For centuries the Sand Lizard people maintained these gardens across the Colorado as "the invaders feared the desert beyond the river," but with gold fever and the region's settlement, the Sand Lizard people and neighboring tribes were killed, arrested, and torn from their land (16). As Grandma and the girls hide from federal authorities following their raid on the Ghost Dance, she teaches them the knowledge necessary so they can survive when she is gone. She describes how to cope with the extreme heat and procure fresh meat, how runoff works its way through the dunes, and where best to plant crops (45–47). When it begins to rain, "Grandma Fleet greet[s] the clouds with tears in her eyes; their beloved ancestors returned to them as precious rain" (47–48). For Grandma Fleet, the elements are an extension of family relations, and standing in the rain brings a deep reverence and closeness to this living and saving resource.

The violence and dispossession described early in the novel surface later through Sister Salt's experience living in Parker, Arizona, on the Colorado River Indian Reservation. Unlike the garden terraces she had cultivated with Indigo and Grandma Fleet, Sister Salt finds the reservation repulsive and oppressive. As "most of the land was above the fertile river bottom, on old floodplains impossible to irrigate," Sister Salt resigns herself to claim land that is too far from the river to irrigate and too parched by the sun to farm successfully. Because "white farmers claimed the best river bottom land," Sister Salt observes the injustices she and the other reservation Indians must live with (204). Removed from their traditional lands and ways of life, the people find themselves in a new world and dependent on the Indian agency and the weekly deliveries of "lard, cornmeal, salt, and a little sugar." For many forced to live on the reservation, they spend their days gambling, drinking, and fighting (205).

This alienation and despair are further perpetuated by the proposed dam in Parker Canyon and a canal to transport water to Los Angeles, corollaries to today's Parker Dam and Colorado River Aqueduct.[11] Sister Salt's relocation to the dam site gives her an intimate view of the land's and river's transformation. Walking with her infant son, Sister Salt sees the growing dam, which "resembled more and more one of the monster stories" her friends had learned as "this monster ate up all living things up and down the poor river" (364). When Indigo, Hattie, and Edward later pass by the dam site on their return from Europe, Indigo observes that the river "was trapped, and only a narrow stream, muddy red, flowed south. The river was stripped naked; all its willows and tamarisks were gone, its red clay banks scraped; and exposed piles of white skeletons of cottonwood trees dotted the swaths of scraped red earth.... [T]he cottonwoods and willows were dying of thirst because the flow of the river was so meager" (394). This destruction of the river corridor not surprisingly mirrors the devastation the tribes experience through the novel as they are uprooted from their homelands and forced within the confines of the reservation. In both cases, the river and the people stagnate.

During the dam's construction, Sister Salt worked as a laundress at the dam site. Yet with the dam's completion, she and Indigo, now reunited, face relocation once again as the rising waters threaten to flood their reservation home. To cope, the girls "made jokes about the river, the government's plan to drown all the Indians, and they laughed and laughed until tears filled their eyes. The only good land left to them now was about to be taken away by the backwater of the dam" (433).

Notwithstanding this potentially tragic end, the novel does not conclude with the girls' destruction, which would reinforce the "doomed Indian" trope Cook-Lynn addresses in this chapter's epigraph. While they have faced years of violence, shame, and loss, they nonetheless find solace in the end, while Hattie, the white woman who attempts to care for and educate Indigo, is the one who "no longer had a life to return to" following Edward's death and her near-fatal beating and rape while on her way to Road's End, where Indigo and Sister Salt live (439). Yet for Indigo and Sister, they return to their beloved terrace gardens. Although "strangers" destroyed the gardens with their apricot trees and the old rattlesnake the previous year, new trees are growing and a rattlesnake has appeared once again (476). Returning the girls to the sacred spring of their people's emergence and noting the devastation caused by settlers' disrespect for this fragile desert ecosystem, Silko leaves her readers to ponder the relationships her Indigenous characters nurture with one another and with their homelands throughout the novel despite the continual racism, abuse, and other obstacles they face. In doing so, Silko creates a work that joins other Indigenous literatures in their representation of self-determination and a continually developing Indigenous presence and future.

Whereas Silko's novels posit a more historical assessment of Indigenous environmental injustice through the broken systems of responsibility at the government's hands and the women's efforts to reestablish those connections through the retelling of stories and returning to their homeland, Tohono O'odham poet Ofelia Zepeda writes of her people's contemporary relationship to their desert homeland and the significance water plays within her culture. In her edited collection *When It Rains* (1982), Zepeda's poems reinforce her people's intimate relationship with the rain. In "Ḍ'ac 'O'odham," or "We Are Papago," the poet ends each of the four stanzas with a line speaking to the process of waiting and anticipating the rain. As they watch an approaching storm, they smell the coming rain, the heat giving way to "a breeze [that] is refreshingly cool." These changes and the imminent prospect of rain make the people "happy as we wait for the rain." However, the storm never materializes. The happiness is gone, and the people feel betrayed by the "lie" the building clouds have told. The poem concludes with a near repetition of the first stanza's final line that underscores the people's identity and its connection to rain: "We are the Papago and se [*sic*] sit here and wait for the rain" (75). For some, this life-sustaining event of a rainstorm in the desert would hardly garner attention, yet for a people whose entire livelihoods revolve around the storms and what they do for the community's crops, such an event merits great attention and respect. Despite the disappointment expressed within the poem through the clouds' lie, the people exercise great patience. This waiting is a symbol of the people, a characteristic of those reared in the desert and intimately aware of the difference any water will make.

Zepeda develops her reflections on water and rain more extensively in *Ocean Power: Poems from the Desert* (1995), which focuses on the daily lives of the women of her tribe and their experiences on the reservation along the Arizona-Mexico border near Organ Pipe National Monument—lands over which Childs travels, as discussed in chapter 2. Although located far from the Colorado River, the reservation lies within the watershed, and, like so many other Indigenous communities throughout the region, the Tohono O'odham rely on Colorado River deliveries along with groundwater pumping and intermittent rains to sustain life.[12] In the introduction to *Ocean Power*, Zepeda dedicates the poems to women who, "like the people before them, . . . gauged the movement of the summer sun and the amount of work that needed to be done. . . . The women planned their day around the heat and the coolness of the summer day. They knew the climate and felt confident in it. They knew the weather and its movements" (1–2). Reminiscent of Grandma Fleet's affinity for place, the poems throughout the collection depict the various relationships Zepeda's family and her fellow community members have with the Sonoran Desert and their reliance on rain and irrigation to survive in one of the nation's most arid regions. These systems of responsibility join Zepeda's extensive mingling of English and Tohono O'odham throughout the collection. Angelica Lawson reads this rhetorical choice to leave the segments of Zepeda's native language untranslated as "a political statement regarding the legitimacy and vitality of the O'odham language—a statement that speaks to the resilience of the language and a resistance to Native language eradication" (181). *Ocean Power*, therefore, reflects Zepeda's efforts to decolonize textual representations of the arid West and assert her own sure voice as a significant contributor to this body of work.

For Zepeda and her people, water is a gift, one cherished and celebrated through both the day's mundane tasks and more sacred rituals. Zepeda speaks to this intimate relationship with rain and water more generally throughout the introduction as she describes how those to whom she dedicates the collection perceive and react to these precious resources. As the women closely follow the weather and the clouds for what moisture they might provide to the parched desert, the women "often laughed out loud" or "laughed in childlike manner" when thunderstorms either failed to live up to their threatening guise or forcefully arrived, soaking the desert and those caught in the storms. Zepeda continues to note how in every instance of rain and its effects on the desert and the people, rain brings utter joy to these women, even in what many might consider the most insignificant details about mud and "the way it coagulated between the children's toes." She also describes how other senses beyond sight react to the rain, how the unique smells of the desert come alive as the rains arrive. "The odor of wet dirt," "the dry bark of mesquite and other acacias," and "the dust that settles on the fine leaves of ocotillo and other leafed plants" create a special bouquet of desert smells that "are all breathed in deeply" by her people (2). Such small details speak to Zepeda's deep awareness of her surroundings and the changes that come with the anticipated storms. For the women in her family, "there was beauty in all these events" that caused them to "[laugh] with joy at all of it." At the same time, however, Zepeda notes how the summer's thunderstorms and their potential for destruction elicit "fear and respect" in the women, another aspect of the relationship these women have forged with the elements over a lifetime and through the stories passed down among the generations (3).

As Zepeda concludes her introductory remarks, she further underscores the sacred place rain has for the Tohono O'odham Nation. In a one-line phrase standing on its own amid the longer paragraphs of this section, she writes, "The rain breaks the tension for the desert. Relief. Cycles continue" (4). This particular phrasing implies the connection the desert has with the rain—that its extreme aridity and summer heat give way to the rain that refreshes the land and its inhabitants, even if only momentarily. As the summer monsoons arrive in the American Southwest, the seasonal cycle continues. The perpetuation of this process depends, for Zepeda, on "those who participated in the ritual of pulling down the clouds and fixing the earth" (5). These include her ancestors, who for generations have petitioned for the seasonal rains to wet the earth and bring forth life as they have since time immemorial. Such details of the people's sensorial understanding of rain, their dependence on it, and the joy and reverence they have for this natural phenomenon embody the system of responsibility Zepeda and the larger Tohono O'odham Nation have for their lands and the waters that support this desert people.

This connection between people and rain defines *Ocean Power*'s first section: "ʾI-huḍiñ g Cewagĭ," or "Pulling Down the Clouds." In the first stanza of the titular poem, Zepeda describes in Tohono O'odham and then English the process of bringing rain to earth. Her narrator uses a "harvesting stick" to "hook the clouds," after which she will bring them down to the earth and "stir" them (9). As the man sleeps, the woman brings forth rain and the man smells the wet earth, dreaming of other women with their sticks in the sky, continuing the process of bringing the rains to the people (9–10). The ritualistic quality of this first poem continues into the next, titled "O'odham Dances." Zepeda writes in Tohono O'odham for the first twelve lines and then transitions to English, where she speaks of the ritual to bring the rain "so that the earth may be fixed one more time," a line that reaffirms again the cyclic nature of the people's petition for rain described in the collection's introduction (12).

As part of the ritual, the people call upon various entities—night, stars, hot breezes, heat, and all animals on land, air, and water—to "focus on the moisture." This is done, as the poem instructs the reader, because "all are dependent" on the water that will mean the difference between life and death (12). Zepeda notes this powerful acknowledgment of all life's reliance on water within the poem's final stanza. Using the collective "we" to emphasize the community's relationship to one another and to their environment, the poet speaks of "humility" as the people face the rising sun: "We are mere beings / All we can do is extend our hands toward the first light" (13). These powerful lines speak to the people's recognition of their place within the world they inhabit. Like the other creatures previously implored, the Tohono O'odham are equally indebted to rain's life-sustaining force. As "mere beings," they acknowledge that all they can do is plead for the rain and then await its arrival. There is no sense of entitlement or desire to use this gift in ways that would compromise other entities' rights to the water—no commodification of water as depicted throughout *Gardens*. The recurring use of "we" and "all" throughout the poem reminds the reader of the collective enterprise that is life in the desert. Herein Gilio-Whitaker's "relationships of reciprocity" take center stage as the poem's human actors purposefully consider and care for (and even rely upon) their nonhuman neighbors.

To conclude this consideration of Zepeda's poetry and its utility in outlining her ethics of reciprocity, I turn to two final poems, one from the first section about rain and the second from a section about irrigation ditches and their role within Tohono O'odham life. Whereas "Pulling Down the Clouds" resonates with the joy and hope the people have through ritual, "The Floods of 1993 and Others" and "The Man Who Drowned in the Irrigation Ditch" speak to the fear and death that are also part of their particular system of relationships. In "The Floods," Zepeda describes the destruction from a flood: "Old trees uprooted / grasses, twigs, and branches all forced" and "Barrel cactus / hanging in uncactuslike manner, / upside down in between tree trunks and large branches." These phrases recall Childs's observations of the flash floods he witnessed in his desert treks. However, Zepeda's representations push beyond his descriptions to give voice to the desert in a more literal manner. "They silently scream," she writes of these plants torn from their moorings through the violence of the rushing water and debris. Perceiving how they must have felt as they were uprooted, the poet then includes the plants' response to her observations: "'My roots are still good, put me in the rocky soil" (21). Here nature petitions the poet for help and, in so doing, demonstrates Zepeda's relationship to her reservation's flora. These are fellow citizens of her world who live and suffer as she does. Not all can hear the organisms' pleas, however. In the third stanza she writes:

> the Park Service employees don't hear them.
> Or if they do, they ignore them.
> Too busy repairing concrete. (22)

Zepeda's portrayal of these employees, likely not of the Tohono O'odham Nation, suggests that either they are unable to hear the screams of nature because they lack the eyes and ears to truly understand what is occurring around them or, if they do hear, their cares lie in attending to the inanimate concrete as part of their work duties.

Zepeda reinforces the connections she establishes between her people and their homes in the collection's longest selection: "The Man Who Drowned in the Irrigation Ditch." Here the narrator tells the story of an elderly man's death in an irrigation ditch and his family's and the community's response to the tragedy. The poem opens with the man's wife and her concern for her husband whose "legs were not strong enough to be straddling ditches" (29). Her imaginations of what accidents might befall him in the community's ditches "where children played, / and ladies sometimes sat and dipped their feet" worry her, for "she knew he would drown, / she knew it was bound to happen sometime" (29–30). The narrator moves from these opening details about the woman's anxiety to a description of the landscape and the "flat, green fields appearing to the foot of distant mountains" that stretched "as far as the eye could see." Establishing the community's dependence on irrigated agriculture from "water pulled from its secret storage place / beneath the earth's surface," the narrator speaks of her people who have farmed these lands for generations, long before federal involvement and the creation of the reservation (30). The poet provides this history, explaining in the third stanza:

> We are called "the people of the cotton fields"
> because of the labor our families did.
> For us there was no reservation, no Housing & Urban
>   Development, no tribal
> support.
> We were a people segregated in row houses
> all lined up along the roads of our labor. (30)

These lines provide important insight into the community's relationship to the local environment and the cultural shifts experienced through colonization of their traditional homelands. Zepeda underscores her people's long residence on the land and their use of water to sustain themselves. Her ongoing use of Tohono O'odham, for entire poems in some cases, juxtaposes the English to demonstrate the two worlds in which she and her people live and her agency in speaking her native tongue to reinforce the age-old relationships with desert water that have sustained her people for countless generations.

The rhetorical sovereignty imbued in Zepeda's work is representative of some of the most recent Indigenous poetry about the Colorado River. In 2020 Mohave poet Natalie Diaz released *Postcolonial Love Poem*, which includes "How the Milky Way Was Made" and "The First Water Is the Body."[13] Both of these poems reflect Diaz's powerful connection to the river, her Mohave and Spanish heritage, and the role that rich, lyric language plays in drawing attention to issues of environmental and identity politics. In "How the Milky Way Was Made," Diaz begins the poem noting the Colorado's historic qualities: It was "once unseparated," a "Red-fast flood" that flowed "all the way to Mexico." She then transitions to the present where the river is "shattered by fifteen dams" and where "pipes and pumps filling / swimming pools and sprinklers / in Los Angeles and Las Vegas" have broken the river (61). In "The First Water Is the Body," this broken river is not only "the most endangered river in the United States" but "also, it is a part of my body." For Diaz, the Colorado is not only the life-giving source for her homelands and people but the essence of her identity. "I carry a river. It is who I am: 'Aha Makav," she writes in "The First Water Is the Body" (46). Recognizing that her readers might misunderstand her meaning in this statement, she emphatically declares:

> This is not metaphor.
> When a Mohave says, *Inyech 'Aha Makavch ithuum*,
> we are saying our name.
> We are telling a story of our existence. *The river runs through the middle of my body*. (46)

Throughout the rest of this four-and-a-half-page narrative poem, Diaz explores the contours of language and her relationship to the river and to her people to establish an integral relationship between the Colorado, the lands and bodies through which it flows, and the Mohave language and beliefs that articulate her respective cosmovision.

Diaz's linkage between these various entities is evident through her depiction of the river's fish, which hold a key place within Mohave cosmology. In light of the river's desecration, she speaks of her people's efforts in the opening lines of "How the Milky Way Was Made" to "save our fish" by removing them from the "skeletoned river beds" and releasing them into the sky, where they became the stars of the Milky Way. Importantly, as she names the particular species of fish sent heavenward, she identifies them by both Mohave and English names: "'Achii 'ahan, Mojave salmon, / Colorado pike minnow" (61). Diaz depicts their movement through the sky, using Mohave once again to connect the fish to the cosmos. "'Achii 'ahan nyuunye— / our words for *Milky Way*" (62). In this two-line passage, Diaz not only links the Colorado via the fish and her body to the cosmos—evident in her description of the connection between body and river in "In the First Body"—but also uses stylistic markers to underscore the colonial impact on her language. Whereas non-English words are typically italicized to denote their foreignness, in this case, Diaz utilizes italics to denote the newness and intrusion of "Milky Way" upon Mohave identity, for such descriptive terms of the night sky fail to capture the Mohaves' depiction of the fish who move "rippling the sapphired sky-water into a galaxy road" (61). Much more could be said about these two poems and the ways in which Diaz moves between languages, cultures, and bodies both human and nonhuman to animate her relationship to the Colorado and all of its relations. Nonetheless, this brief reading further underscores the significant contributions Indigenous writers and poets are making to the knowledge of the river and its peoples. Like Silko and Zepeda, Diaz clarifies her people's relationality to the river, underscores its transformations and the impacts these have had upon the Colorado human-nature interface, and witnesses a commanding Indigenous presence within the watershed whose past, present, and future are intimately connected to the health of the river. As Diaz writes, "a river is a body, as alive as you or I" (48).

Ten Tribes Partnership Tribal Water Study

The representations of trauma, reciprocity, resistance, and resilience depicted in the previous texts powerfully capture these writers' relationships to place and the colonial history that has shaped tribal associations with the natural world and with one another. Yet as Indigenous communities throughout the basin have begun to assert their reserved rights, they manipulate narrative within other genres and alongside instrumental, scientific, and rhetorical discourses to reinforce the presence and consideration of their unique water cultures. One of the most extensive and recent collections of Indigenous voices comes from the Colorado River Basin Tribes Partnership, or Ten Tribes Partnership, formed in 1992 to "claim their seat at the table and raise their voices in the management of the Colorado River as water challenges persist" ("The Ten Tribes Partnership").[14] This partnership includes the Ute Indian Tribe of the Uintah and Ouray Reservation, the Ute Mountain Ute Tribe, the Southern Ute Tribe, the Jicarilla Apache Nation, and the Navajo Nation from the Upper Basin. The Fort Mojave Indian Tribe, the Chemehuevi Indian Tribe, the Colorado River Indian Tribes, the Quechan Indian Tribe, and the Cocopah Indian Tribe constitute those from the Lower Basin. Together, they represent a powerful force for promoting Indigenous rights throughout the Colorado River Basin, as they hold claims to 2.8 million acre-feet of Colorado River water (*Colorado River Basin Ten Tribes* 1-1). The partnership's study represents a significant development of the tribes' rhetorical and hydrological sovereignty.[15]

In December 2018, the partnership joined the US Department of the Interior and the Bureau of Reclamation to publish the extensive *Colorado River Basin Ten Tribes Partnership Tribal Water Study*, outlining each of the tribal nations' relationships to water and past, current, and future water needs. Informed by the 2012 Colorado Basin Supply and Demand Study, which outlines the basin's future water needs as a whole, the partnership's *Water Study* ensures that many of the region's tribes articulate "from their own perspective" what they see as the basin's water issues (i). The 362-page report begins with the foreword authored by both the commissioner of the Bureau of Reclamation, Brenda Burman, and the Ten Tribes Partnership to reflect the unique legal relationship that unites the US government with these tribal nations as sovereign entities. This opening and the rest of the extensive document reflect the partnership's efforts to "control the story" of Colorado River use by the basin's Indigenous communities, a move Timothy Casey identifies as an important strategy in shaping policy (161). The partnership's foreword underscores from the beginning that "water is life" and situates findings about tribal water interests within the watershed's broader context. That is, the partnership's goals are to better understand how their current respective water use fits into the basin's larger management picture, how tribal development of reserved water will impact entities using tribal water, and how future tribal water development will impact the basin going forward (*Colorado River Basin*

Ten Tribes iii). Such goals demonstrate just how significant tribal water rights have become within the watershed. Where they historically have had minimal impact on the basin's overall management, today they have become the "slumbering Monstro of the Southwest"—a juggernaut that will have a significant impact on future Colorado River water development (J. L. Powell 154). The partnership's *Water Study* and the articulations of each respective tribe's water rights and unresolved claims cannot be denied. Future deliberations will have to acknowledge these rights established more than a century ago with the *Winters* case.

At first glance, this comprehensive *Water Study* resembles a typical hydrological report filled with scientific data and instrumental discourse to guide management scenarios. However, narratives reflective of the tribes' many connections to the river punctuate the document in a manner reminiscent of AEURHYC's calls for delta restoration. Thus, with these narratives infusing much of the report, the *Water Study* takes on a very different guise than the run-of-the-mill reports published by the bureau. Indeed, tribes engage narrative to evoke their cultural and spiritual values, "Indigenous scientific literacies," which Adamson and Monani reveal are "complex multispecies entanglements that imaginatively argue for a safer, livable present and future" (14).[16] For example, evidence for such literacies appears in the partnership's water-is-life appeal, which is far more than an obvious statement about the element's fundamental nature to all life. Rather, it is a refrain expressing deep cultural and spiritual values that define the tribal nations' stewardship of this material entity.

The partnership's introductory vision statement captures this broader understanding of water's significance within an Indigenous context. The statement reads, "Water is life. Water is the giver and sustainer of life. Water is a sacred and spiritual element to the Tribes of the Partnership. The Creator instilled in the First Peoples the responsibility of protecting the delicate, beautiful balance of Mother Earth for the benefit of all living creatures. The Partnership will embrace and own the stewardship of the Colorado River and lead from a spiritual mandate to ensure that this sacred water will always be protected, available and sufficient" (*Colorado River Basin Ten Tribes* 1-1). The repeated emphasis on the spiritual relationship to water and the tribes' sacred duty to protect and maintain this essential resource draws a noteworthy difference from how traditional, non-Indigenous water reports articulate the importance of water. The partnership establishes its rhetorical and hydrological sovereignty by clearly stating its role in the Colorado's future: It will eagerly oversee the river's caretaking and will be the leaders on this front since many other entities, including the federal government with its legal responsibility to the tribes, have failed to adequately protect the river's waters for all stakeholders. The Ten Tribes Partnership's official website affirms this direct and deliberate language about the partnership's mission as it opens with a panoramic image of the Colorado River winding through desert canyons with the title "Keepers of the River" and a selection from the *Water Study*'s vision statement emphasizing the partnership's role in stewarding the river ("Ten Tribes Partnership"). Significantly, this position also infuses river management with spirituality. Nowhere does this passage set scientific knowledge as the de facto epistemology by which to direct the river's future. Instead, the *Water Study* boldly articulates the partnership's cosmovisions by which the tribes will govern their allocations—approaches that stand in stark contrast to mainstream river governance.

The *Water Study* emphasizes this unique connection between the cultural and spiritual relationships that defines their stewardship particularly in chapter 5, which offers an extensive overview of each of the ten tribes and their historical connections to the Colorado and its tributaries. For example, the section on the Ute Mountain Ute Tribe outlines their historic territory that ranged from southern Colorado and into Utah and New Mexico depending on the time of year and their eventual relocation to the reservation in southwest Colorado where they turned to ranching and agriculture (*Colorado River Basin Ten Tribes* 5.3-3). The tribe also notes the inadequate water infrastructure and the nonirrigable lands that characterize much of the reservation (5.3-3–4). This brief history reminds the audience of the Ute's long-standing tenure in the area and the challenges they have faced since relocation as they have had to adapt to settler-colonial lifestyles. Despite the upheaval that has changed their traditional ways of life and connections to historic homelands and waters, the Ute Mountain Ute continue to maintain important cultural and spiritual relationships with the waters around them. The tribe explains, "Water brings life, sustenance, and is a tool of blessing and prayer for the Tribe," particularly during the Bear and Sun Dances that renew tribal identity and meaning (5.3-4). By outlining these sacred connections to water, the tribe establishes the rationale for subsequent discussions about its current water supply and the unresolved water claims it has with New Mexico and Utah.

The *Water Study*'s most extensive discussion of a system of reciprocal responsibilities regarding the basin's waters and inhabitants along cultural and spiritual values derives from the segment by New Mexico's Jicarilla Apache Nation. Following brief information about the nation's reservation, the document provides a lengthy overview of their historic conditions to water. The nation states, "Water, in all its forms . . . is sacred to the Jicarilla people and has been a fundamental tenet of the Nation's religion since time immemorial. In Jicarilla creation stories, 'Born of Water' is a major deity who made the world a safe place for human habitation and brought the four sacred rivers—the Arkansas, Rio Grande, Canadian, and Pecos Rivers—to the Jicarilla people" (5.4-3). This explanation of water's significance to the Jicarilla's cultural identity establishes the permanence of this spiritual relationship between the people and the water, one that is "to be protected and honored through religious ceremonies and rituals" (5.4-3). Thus, water has been and continues to be a key component of the Jicarilla Apache's cosmology and identity as the sacred rivers delineate their homelands that too are "sacred" and that define "Jicarilla religion, culture, lifestyles, and their very identities as a distinct people" (5.4-3–5.4-4). As such, water "has never been just a commodity or a necessity of life" for the Jicarilla Apache Nation, "but a sacred element that requires respect, reverential treatment, and efficient use" (5.4-4).

Evident throughout the Jicarilla's statements is the role water has played in the past and continues to perform today. This connectivity underscores the nation's rights to water established long before Anglo settlement of their homelands and the need today for the nation to develop economically while maintaining this sacred relationship. The nation concludes this section of its respective water study reminding readers, "The reverence for and appreciation of the scarcity of water continue to dictate the Jicarilla's individual and cultural relationship to their homeland. The features of the landscape, especially its water resources, are instrumental and integral to the nation's modern economic development and the preservation of this ancient culture in the 21st century" (5.4-4). These declarations about their sacred connections to the rivers and their homelands, and how these connections are vital to the tribe's future prosperity economically and culturally, assert their fundamental rights to this entity based on the "first in time" principle. Such statements powerfully challenge notions of the vanished or doomed Indian myths that too often inform modern America's understanding of Indigenous cultures.

For some of the tribes in the Lower Basin, ongoing environmental injustices frame their responses to Colorado River management. The Fort Yuma–Quechan Tribes note the impacts from the dams upriver that have cut off flood events and the natural deposition of sediment that once enriched Quechan agriculture within the Colorado's floodplains (5.9-5). The challenges are more pointedly expressed by the Cocopah Tribe, who have suffered much due to the dramatic disappearance of the Colorado's rich delta ecosystem. The Cocopah describe their historical uses of the river and then point directly to how that life "all changed with the coming of the European settlers." Citing the US-Mexico boundary disputes and treaties that eventually separated the tribe between the two countries, the tribe explains that the Colorado "was eventually taken from them and the Tribe was left with dirt." The Cocopah then ask the audience a number of rhetorical questions about the fate of the tribe, concluding their statement about the historical and cultural uses of the river with a powerful final statement blatantly calling out the injustices imposed on them by the colonial system: "Cocopah (Kwapa) are known as the river people, but there is no river left. They have had a centuries old way of life taken from them" (5.10-4). Repeating the historical connections to the river as other partnership tribes have done, the Cocopah remind readers that despite these pervasive environmental injustices, they are still there fighting for their rights to the river and traditional ways of life. Such efforts demonstrate the *Water Study*'s significant status as an artifact of Indigenous sovereignty as the partnership articulates in its own words and on its own terms how the Colorado River has been used to threaten tribal sovereignty and how the partnership is taking control of its own destiny today.

As an important tool in asserting their rhetorical and hydrological sovereignty, each tribal nation's statement on the cultural, spiritual, and economic uses of water upends historically inaccurate ideas about the region's Indigenous presence. These statements, expressed often in narrative form to delineate tribal histories and, at times, creation stories, as in the case of the Jicarilla Apache, combine with extensive commentary on storage capacity, flow rates, irrigation and groundwater infrastructure, and projections of water-use scenarios for each of the partnership's tribes. The *Water Study*'s discursive hybridity demonstrates the partnership's rhetorical skill in engaging with the dominant scientific and instrumental discourses that shape water management while at the same time incorporating statements reflective of their ethical and moral standing on water's being and connection to the partnership's various identities.

The partnership reinforces this deft use of these various discourses through the numerous photographs throughout the document, which also reinforce the historical rights the tribes have to the water, their advancements over the past 150 years or so to develop these rights, and the challenging conditions that prevent Indian water use from matching the efficiency and range of access evident in non-Indigenous communities. In the case of the Fort Mohave Indian Tribe along the Arizona-California border, the images they provide showcase the progress and self-sufficiency the tribe has made through its reserved rights. With a diversion right to 132,789 acre-feet from Arizona, California, and Nevada, the tribe has constructed a golf course as well as a hotel and casino and also manages AKA Farms, which produces alfalfa and other crops (5.6-4–5). Photographs of palm trees and verdant lawns, as well as thriving fields that include a section of eighty thousand olive trees irrigated by a center-pivot sprinkler system, which the tribe notes is the "first of its kind in the world," demonstrate the tribe's assets and technological savvy, making them an important agricultural voice in the region (5.6-6). These particular representations also work to underscore the inequities of tribal water systems evident in the Southern Ute Indian Tribe's image of a 1927 diversion gate that is still in use "despite its antiquated condition," as the caption reads (5.2-5). This discursive manipulation as a form of rhetorical sovereignty or Gilio-Whitaker's "native resistance" can also be expressed as "engaged resistance," Dean Rader's term that Adamson and Monani summarize as "a fundamentally Indigenous form of aesthetic discourse that engages both Native and Western means as resistance against," in Rader's words, "colonial assimilation and erasure" (qtd. in Adamson and Monani 15). These strategic moves throughout the *Water Study* reflect the partnership's keen awareness of the high stakes involved in securing their legally allotted water and stewarding its care by underscoring the river's sacred nature for all river users now and in the future.

For many individuals and communities throughout the Colorado River watershed that have relied on Indigenous waters in the past to support their respective needs, the partnership's *Water Study* might raise fears about the tribes' growing strength as a powerful water broker. As if to placate such sentiments, the partnership's foreword notes that the potential water-use scenarios revealed that while tribes will look to develop their federally reserved water rights, they are not the real concern. "If there is a 'take-away' that was surprising," the partnership writes, "it is that, even under the most favorable of circumstances for rapid tribal water development, the amount of water that will be used by the tribes is dramatically overshadowed by the effect of climatic conditions on the overall supply of water in the basin. Nature is still in charge" (iii). The partnership's awareness and acknowledgment of nature's power and influence on the future directions of Colorado River management are crucial. The real threat is not the tribes, but persistent drought and the failure to live within the environmental confines nature dictates and will prescribe in the future. While tribes throughout the basin will use water on a larger scale than before, their deep generational knowledge of living with aridity and human-caused shortages makes them the obvious choice to lead as the river's keepers in guiding the rest of the basin's inhabitants to live within a smaller water budget through greater humility, restraint, and appreciation for the value the river has for all.

"The Voices of Grand Canyon" Digital Project

Just as federal agencies such as the Bureau of Reclamation have begun to listen to and engage tribal nations more closely on water resource issues, so too have environmental organizations recognized, albeit long overdue, the value of Indigenous leadership and participation in shaping the public's understanding of the river and watershed. The Grand Canyon Trust's "The Voices of Grand Canyon" campaign is one such initiative to foreground Indigenous systems of responsibilities in this particular region. This nonprofit, which focuses on environmental and cultural issues throughout the Colorado Plateau bioregion, released this collection of responses on its website in February 2020. The voices include Jim Enote (Zuni), Nikki Cooley (Diné), Leigh Kuwanwisiwma (Hopi), Coleen Kaska (Havasupai), and Loretta Jackson-Kelly (Hualapai), who recount their individual and tribal connections to the Grand Canyon and the Colorado River. Their extension of the oral tradition through digital media, comprising brief videos and print statements, combines to reinforce the tribes' historic connections, which assert their "first in time" claims to the river and its surrounding lands.

Similar to the expressions of cultural and spiritual relationships within the *Water Study*, Jim Enote, the chief executive officer of the Colorado Plateau Foundation, speaks of his tribe's origins in the Grand Canyon and the petroglyphs on Zuni lands to the east and within the Grand Canyon to reaffirm the people's deep connection to these sacred places within the

"The Voices of Grand Canyon." Photo by Deidra Peaches, design by Stephanie Smith, Grand Canyon Trust.

larger Colorado Basin. For the Zuni, these stories "are telling us we should never forget where we came from."[17] Noting the Hopi's particular emergence within the Grand Canyon, Enote provides a history of the Zuni's origins at Chimik'yana'kya dey'a, or Ribbon Falls, explaining that they lived in the canyon for "a long, long time" and then "emerged, exploring all the tributaries of the Colorado River" before uniting on the lands that presently constitute Zuni Pueblo. While petroglyphs carved into rock long ago capture this movement from the canyon and beyond, the performance of Zuni rituals continues to reinforce their ancestral home. The perpetuation of this knowledge helps Zuni and outsiders understand the tribe's deep ties to this sacred space.

The Grand Canyon's significance as a place of emergence also defines Leigh Kuwanwisiwma's perspectives on the Hopi's affiliation to the area. But for the Hopi, the Grand Canyon is not just the tribe's "genesis" but their "final spiritual home," as their spirits will return to the confluence of the Colorado and Little Colorado Rivers to then take on the form of clouds, which will then circle the earth. Contrasting popular beliefs about the Grand Canyon's supernal wilderness qualities, which tend to neglect the canyon's human presence, Kuwanwisiwma asks that visitors to the area understand that "the Grand Canyon and humanity are all one." Coleen Kaska similarly invites the canyon's visitors to consider the region's Indigenous history. She contemplates the Havasupai's loss of tribal lands and how Grand Canyon National Park's centennial celebration in 2019 represents for her a century of dispossession and the failure of the government to fulfill its obligations to the Havasupai by recognizing their claims to their traditional lands, including their emergence place at Red Butte. Raised on the stories of her people's ancestral connections to the Grand Canyon, Kaska directly addresses the audience, reminding listeners that "Natives are still around.... And they will never forget. I will never forget." For Kaska, like Kuwanwisiwma, the canyon is a powerful reminder of her people's past and the injustices that have characterized the tribe's relationship to the canyon, particularly after the establishment of the Havasupai reservation, which greatly reduced their use of and access to their traditional homelands.

As one listens to and reads these diverse tribal cosmovisions, it is clear that the narratives used therein operate beyond a simple informative overview of each tribe's historic connections to the Colorado River and the Grand Canyon. Instead, these discussions about sacred origins emphasize the need to bring attention to current issues of sovereignty and survival the tribes face in light of ongoing threats to their lands and tribal identities. Loretta Jackson-Kelly describes the significance of the middle of the Colorado River as "the backbone of the [Hualapai] people" and that "without that backbone, we cannot survive"—a point shared in Diaz's poetry about the Mohave connection to the Colorado. Jackson-Kelly's work in the tribe's cultural resources department and tourism agency ensures that she is able to teach this essential relationship between her people and the river, whose native fish are her people's ancestors, to the Hualapai Reservation's many visitors. She notes that many outsiders come to the reservation with incorrect ideas about Indigenous peoples and that she wishes they would "try to understand the world that they're going to step into" so as to recognize the ongoing presence and vitality of the Hualapai and other Indigenous peoples.

This ongoing effort to dispel negative stereotypes and instruct non-Indian peoples about Indigenous survival through the perpetuation of their cultural and spiritual relationships to the Colorado and Grand Canyon highlights Nikki Cooley's contribution to "The Voices of Grand Canyon" project. Like Jackson-Kelly, Cooley seeks to dismantle myths about the West and Indigenous peoples perpetuated by popular culture. Her unique position as the first Diné woman licensed as a Colorado River guide allows her to challenge her customers' oft-held beliefs that "Native American culture...was lore. It was John Wayne movies." Such on-the-river experiences with misinformed tourists reiterate the importance of acknowledging the very real violence her people faced, how the Grand Canyon was a refuge for them during the time of relocation, and how it continues to shape the practices and beliefs of a people who are very much involved in shaping the Southwest's present and future. "It's a place of resilience in more than one sense and can teach us about the history of a people who were trying to survive—the people who lived, persevered, and are still here today," she explains.

Informing her fellow rafters of this intimate connection, Cooley also invites them to see the river and canyon beyond the sublime recreational appeal for which they are so highly regarded by visitors across the globe. The canyon "is not a museum to be gawked at," as she explains, but a place of inestimable sacred value she equates to her "church." She emphasizes this point through her elaboration on the Diné's view of the Colorado's confluences with its tributaries. Cooley explains, "The big Colorado River is considered the male river, the Little Colorado and the San Juan, they are all considered the female rivers. And where the waters come together, the confluence, together they nourish the rest of the Grand Canyon. It is a very sacred place that we must treat very carefully, respectfully, and not think of it as a theme park" ("Voices"). This last reference to how some view the canyon likely refers to recent development plans for the Colorado and Little Colorado Rivers such as the defeated Escalade Project and the proposed hydroelectric dam on the Little Colorado.[18] As the confluence and cultural sites throughout the Colorado and Little Colorado corridors are sacred to the Diné, Hopi, and other tribes, the construction of these projects would be a direct affront to tribal identity and sovereignty. Cooley and her fellow "Voices of Grand Canyon" contributors represent powerful tributary voices that reject mainstream beliefs separating nature from culture and that forward myopic and, in many cases, superficial perspectives about the river's and canyon's worth. As these five tribal voices attest, the Colorado and its celebrated canyon demand greater reverence and respect, as do the various tribes whose past, present, and future are so intricately aligned with these entities.

The sampling of voices addressed in this chapter and their use and manipulation of written and oral texts demonstrate how various tribal nations within the Colorado River Basin have asserted and continue to exert their rhetorical sovereignty to develop comprehensive Indigenous cosmovisions of and systems of responsibilities to all the basin's entities. These systems reflect the sacred nature of the Colorado, its tributaries, and the lands through which they flow. They honor the many diverse peoples who depend on the rivers for survival as well as the flora and fauna, which animate the basin's life-world and are fellow citizens to the millions of people vying for diminishing resources. Such expressions, articulated through various manipulations of western discursive strategies, highlight the environmental destruction dams have inflicted on tribal lands, as well as the methods that have historically separated tribes from their legally allotted water shares. Coupled with recent events such as those surrounding the Dakota Access Pipeline protests in the Missouri River Basin and Utah's recent quantification settlement with the Navajo Nation, these examples reflect the growing strength of a collective Indigenous movement to take the lead in caring for the nation's water bodies.[19]

However, much work to remedy these injustices remains. While the *Water Study* and "The Voices of Grand Canyon" project are important efforts to move Indigenous voices and the issues they promote into the mainstream, they still reside, all too often, on the periphery. The Navajo Nation's COVID-19 crisis has reinforced this unfortunate fact. In a letter to the *Navajo Times*, the authors, hailing from Gallup, New Mexico, and Window Rock, Arizona, respectively, rail against the economic conditions on the Navajo Reservation and the continued lack of water infrastructure.[20] They note that as families live together in communities where domestic water exists or travel to nearby towns to purchase water, they put themselves at risk of contracting the coronavirus. For them, "The COVID-19 crisis in Navajoland today is partly also a water crisis" (Kelley and Francis). Similarly, Jack Ahasteen's May 21 and 28, 2020, political cartoons for the *Navajo Times* corroborate these sentiments as they depict the economic inequality that exists on the reservation caused by water insecurity. One cartoon, "Control the Water and You Have Everything," depicts two men driving a truck hauling water, looking beyond the reservation's border to a desert oasis where palm trees, a lake covered with watercraft, and high-rises glisten in the summer sun. The men driving the truck, their faces covered by masks, look toward the city and exclaim, "As long as we have water we have some hope!" while the bottom of the cartoon reads, "Control the water and you have everything." The second image, "$600 Million Care Act Water Line," wryly comments on the recent legislation and the reservation's water woes as it depicts a human chain passing buckets of water between individuals masked to prevent the coronavirus's spread. The quips at the bottom of the drawing read: "hire local, buy local" and "unemployment rate 0."

These recent examples highlighting the economic inequities on the reservation reflect the ongoing and very real challenges Indigenous communities throughout the Colorado Basin face as they pursue water security. However, the *Water Study* and "Voices of Grand Canyon" project represent significant steps in asserting the tribes' "first in time, first in right" position as they underscore their long-standing presence in the region and the various relationships they have cultivated with the river over millennia. Bolstered by substantial water rights and the power associated with those rights, the tribes will increasingly become more prominent participants at the bargaining table. Yet unlike many of those who have historically managed the river for purely political or economic aims, the tribes have other values guiding their decisions. From the partnerships established in these projects between the tribes, the Bureau of Reclamation, and the Grand Canyon Trust, there is hope that the values outlined herein by the Keepers of the River can reach a broader audience so that new stories may be told that invite a greater reverence and respect for the Colorado and all its relations.

CHAPTER SIX

Reclaiming Reclamation

Stewarding Water Resources in the Twenty-First Century

The story of irrigation in Utah is the saga that began with the individual irrigator, shovel in hand, coaxing a trickle of water onto the dry land. The history of Utah is indeed inextricably linked to the development of irrigation.
—Craig Fuller, "Irrigation in Utah"

The emphasis Latina/o and Indigenous communities place on the spiritual connections to the river as evidenced in the previous chapters raises important questions about the relationships between spirituality, religion more broadly construed, and caring for the natural world. We might consider whether such beliefs have any place in water deliberations so dictated by appeals to reason, objectivity, and science—positions that all too often seem to run contrary to those espoused by faith and belief. Moreover, if they do make a difference, how might one's belief in a divinely created world promote a more sustainable water ethic for the Colorado River? What are the ideas, principles, or values that exist among various religious traditions that may encourage this alternative ethic? Which religious traditions have significant historical influence within the watershed, and how might their histories and theology influence current attitudes and behaviors about nature, which might ultimately offer valuable insights about how people and the river might coexist in a more harmonious, healthy manner?

In light of this last question, two religious traditions stand out for their impact on shaping the American West and its water resources. The first is Catholicism and its impact on the exploration and settlement of the Southwest through the mission and acequia systems of managing converts and water. Today, the communal acequia, or ditch system, continues to define irrigation efforts in the Hispano communities throughout the Rio Grande Basin, in particular, and in other places such as the San Juan Valley. Sylvia Rodríguez's treatment of these acequia-based communities throughout New Mexico and southern Colorado demonstrates the ways in which religious belief infuses daily water use. To ensure a bounteous crop, processions occur within these communities in the spring wherein devotees carry the carved images of San Isidro Labrador, the patron saint of farmers, or San Juan Bautista, patron saint of water, throughout the towns and along the acequias. In doing so, Rodríguez notes that the processions "resacralize the space through which they pass, an agricultural landscape defined by water sources, ditches, and built structures, including holy sites" (102). In essence, the processions reaffirm the practitioners' commitment to one another both through their water use and through their shared religious devotion. It shores up their religious and geographic identities, reminding them of who they are and the responsibilities they have to subsequent generations. Thus, as Rodríguez further suggests, "Those who perform religious rituals believe that their actions promote harmonious relations among community members and with the divine power that controls nature, life, and death," a powerful articulation once again of the ways in which *querencia*, this time aligned with religious belief, shapes community responses to water (103).

Joining the Indigenous and Hispano influences on water development in the West is the Church of Jesus Christ of Latter-day Saints.[1] Fleeing mob violence and evicted from their homes in a number of eastern and midwestern communities, the Latter-day Saints arrived in the Great Salt Lake Valley in the summer of 1847. There, they established a new home on Ute, Shoshone, Paiute, and many other Indigenous lands throughout present-day Utah and the larger Intermountain West they sought to colonize throughout the nineteenth century.[2] A group whose presence in the West has garnered significant admiration and scorn, for both its religious beliefs and its religious practices, as well as its impacts on the lands and peoples the Latter-day Saints attempted to settle, civilize, and convert, their influence on the region and the Colorado River nonetheless deserves further consideration, as the church, perhaps more than any other group of western settlers, embraced the full tenor of Isaiah's biblical vision: "The wilderness and the solitary place shall be glad... and the desert shall rejoice, and blossom as the rose" (35:1). Finding a refuge in the Great Basin after decades of persecution, the Latter-day Saints went to work to realize the garden myth in ways that would eventually become the envy of other western communities.[3]

My discursive reclamation of this tributary voice has two primary goals: The first is to advocate for the value that religion has in general regarding questions of sustainability. Despite a growing interest in the intersection between religious belief and resource development over the past couple of decades, religion—whether at the institutional level or in terms of doctrine—is all too often absent when addressing environmental challenges, including those about water. Such marginalization ignores both the valuable contributions religious teachings have in constructing ethics of care and the power that religion has in motivating the behavior of billions of the earth's inhabitants based on such ethics. This chapter's second objective is to reclaim reclamation or the particular notion of western reclamation that has governed western water development for more than a century. To enact this reassessment of one of the region's most cherished beliefs within the context of twenty-first-century constraints along the Colorado, I turn to the Church of Jesus Christ of Latter-day Saints' water culture of the nineteenth and early decades of the twentieth centuries, as well as the church's doctrinal teachings about the earth. Here I follow George Handley's thoughts on addressing the climate crisis through the church's doctrine when he writes, "We don't need new teachings but perhaps a clearer and more explicit emphasis on our responsibilities to honor the Creation," wherein "environmental stewardship could be more explicitly underscored when teaching the principles of provident living and self-reliance" (*Hope of Nature* 164). Thus, as with the other tributary voices in this study, the perspectives I examine in this chapter are not new. Rather, the reframing of provident living and stewardship within the context of our current environmental crises gives them new significance. Therefore, as I contend here, the church's early irrigation development with its respective successes and failures, the contributions of John A. Widtsoe (a church leader and one of the most ardent supporters of western reclamation), and the church's teachings regarding stewardship and provident living offer a useful narrative of reinhabitation for members and nonmembers of the church alike to invigorate discussions about the Colorado River. In doing so, these narratives challenge current attitudes and practices to promote a more sustainable reality for this threatened entity. Ultimately, this reassessment of western reclamation through a Latter-day Saint lens invites a more comprehensive, interdisciplinary approach to water issues by reimagining what foundational settler-colonial tenets such as Isaiah's vision mean for the West today and emphasizes how mobilizing institutional beliefs can better the Colorado and the larger environment.

Religion, Ecology, Water

Lynn White Jr.'s 1967 essay "The Historical Roots of Our Ecological Crisis" has had a significant impact on how Western society perceives the relationships between organized religion, and particularly the Judeo-Christian tradition, and nature. In the article, White contends, "Christianity is the most anthropocentric religion the world has seen" and "bears a huge burden of guilt" for the current environmental challenges that the planet faces (1205, 1206). Other scholars have admitted to similar assumptions, as does Max Oelschlaeger when he states, "For most of my adult life I believed, as many environmentalists do, that religion was the primary cause of ecological crisis" (1). Yet like White, who looks to Christianity and Saint Francis as a corrective to the problems he outlines—a perspective many often overlook when citing White's supposed screed against religion—Oelschlaeger ultimately concludes that "religion is a *necessary condition* for the resolution of ecocrisis" (22). Although religion—or spirituality more broadly—has always been part of environmental thought as transcendentalism and the sublime articulate one's direct experience of the divine through nature, not until recently has a clearly demarcated field of religion and ecology emerged to tackle present environmental challenges.

Central to the work of scholars working in the field of religion and ecology is their justification of religion within the realm of environmental crises and the search for commonalities between those who profess a particular faith and those who do not. As the editors of *Grounding Religion* (2011) suggest, "Many people, both within and outside of traditional religious structures, experience a profound sense of something spiritual and holy in this natural world" (Bauman, Bohannon, and O'Brien 2). Roger Gottlieb, editor of the *Oxford Handbook of Religion and Ecology*, further elaborates on this relationship between believer and nonbeliever, stating, "Both believe that life deserves reverence that cannot be reduced to dollar value, that self-examination and spiritual practice make the most important kind of sense, and that there is more to human well-being than money, power, and pleasure" (16). Religions also powerfully contribute, as they are no stranger to restraint and sacrifice. For this reason, they are well equipped to address the environmental challenges that will certainly require people to make changes to their behavior. As Gottlieb confesses in regards to living one's religious beliefs, "People do often make much of how hard this is, ignoring the fact that a committed religious life is always difficult: loving our enemies or giving up attachment to desire is, after all, no day at the beach" (11). For these simple yet profound reasons, religion has much to offer to questions of sustainability. Perhaps the most compelling reason for turning to religion to address our environmental challenges, however, is that the majority of the world's population expresses belief in a higher power and relies on religious belief and practice to give meaning to life and to govern one's ethics. Mary Evelyn Tucker and John Grim, the cofounders of the Yale Forum on Religion and Ecology, argue for a wide coalition of voices to address environmental challenges, including those representing religion "because the attitudes and values that shape people's concepts of nature come primarily from religious worldviews and ethical practices. The moral imperative and value systems of religions are indispensable in mobilizing the sensibilities of people toward preserving the environment for future generations" ("Overview"). As a central clearinghouse for many different religious traditions across the globe and their respective positions on the Creation and climate change, the forum seeks to enact the interdisciplinary conversation that is necessary to tackle pressing environmental problems.

Other scholars and activists have identified the important intersections between religion and water, noting what happens when societies forget or ignore such connections. As Maude Barlow and Tony Clarke write in *Blue Gold* (2002), "At the heart of any new water ethic must be a renewal of our ties with the natural world and a reverence for water's sacred place in it" (211). This reconnection of our relationship to water is vital, as "modern society has lost its reverence for water's sacred place in the cycle of life as well as its centrality in the realm of the spirit, and this loss of reverence has allowed humans to abuse it" (221). One way of reestablishing this link is by reclaiming the sacred link that religions have with water. In his own argument for a new water ethic within the American West, Daniel McCool, a Utah-based water scholar, points out that "even though the academic and philosophical debate over a 'new water ethic' is a contemporary phenomenon, every religion on the planet has long employed the metaphor of water" (229). Such deep history and memory strengthen adherents and offer a pathway by which to reconstruct a more sustainable relationship with water and all life dependent on this entity.

Latter-day Saint Water Culture

The relationship the Church of Jesus Christ of Latter-day Saints has with water runs deep in the arid West. When the first group of wagon trains entered the Salt Lake Valley in July 1847, they found what Utah historian Thomas Alexander describes as the "Wasatch Oasis," a setting of lush grasses and fertile soil watered by numerous mountain streams that blanketed the valley ("Irrigating" 172–73). Although newcomers to the region, the pioneers had discussed irrigation while in Nauvoo, Illinois, prior to their exodus west (Arrington 41). They had also observed irrigation techniques from their interactions with the Native, Hispano, and even Anglo communities through the American Southwest as they traveled westward and began colonizing the region. Members of the Mormon Battalion, formed at the request of the United States to assist its campaign against Mexico, observed acequia culture in Santa Fe and surrounding communities in 1846, and then later in the California gold mines, while Parley P. Pratt, a prominent leader within the church, witnessed small-scale irrigation among the Indians in southern Utah (Worster, *River Running* 352; Harper 37–48; Fuller). Building on these earlier forms of water development, the Latter-day Saints would be the first to implement irrigation throughout the West on a large scale. Elwood Mead, the Bureau of Reclamation commissioner from 1924 to 1936 and the namesake for Hoover Dam's reservoir, declared Utah "the cradle of our modern irrigation history" (42). As is reported in countless histories of the reclamation of the arid West, upon their arrival in the Salt Lake Valley, one of the Latter-day Saints' first tasks was diverting City Creek and planting crops.[4]

Latter-day Saint irrigation spread throughout the West as Brigham Young, president of the church until his death in 1877, directed the faithful to colonize hundreds of sites throughout what Young envisioned as the State of Deseret. Stretching from the Sierra Nevada on the west to Colorado's Rocky Mountains on the east, this latter-day kingdom of God on earth encompassed portions of California, Oregon, Idaho, Wyoming, Colorado, Arizona, and New Mexico. Beginning in 1847, Latter-day

Saints founded more than five hundred colonies in the nineteenth century, including those in Canada and Mexico (Arrington 88). To succeed, these nascent settlements relied on a communal approach to water development reminiscent of the acequia culture in the Hispano communities. Similar to the acequia system's *mayordomo*, a water master oversaw the division of water among those under his watch, and all those using water were expected to provide labor to maintain ditches in proportion to the amount of water they needed to irrigate their crops (Arrington 53). To underscore this communal spirit, Brigham Young supposedly taught the people in 1848, in a now widely referenced statement, "There shall be no private ownership of the streams that come out of the canyons, nor the timber that grows on the hills. These belong to the people; all the people" (qtd. in Arrington 52).[5] This sentiment dictated the Latter-day Saints' approach to natural resources for the next two decades as the church administered Utah Territory's waters to all the people based on their needs, rather than following the doctrine of prior appropriation developing in neighboring states. "Mormons did not think of water rights the way most western immigrants did; for them, water was not viewed as property," as Utah author John Bennion explains ("Water Law" 292). This communal approach lasted until 1897, when the state legislature adopted the prior-appropriation system (295).

The Latter-day Saints' approach to water development, first applied within the Great Basin's valleys on the Wasatch Mountains' western slopes, spread elsewhere throughout Latter-day Saint colonies, including those in the Colorado River Basin. In fact, the church had long shown interest in the Colorado River as an essential component of the church's growth. As early as 1846, the church considered how the Colorado River might aid their colonizing efforts (M. Smith 29). In the 1850s, this interest grew as Young established the Southern Indian Mission, led by Rufus C. Allen, to whom he gave the charge to "explore the river and determine its navigability" (32). When Lieutenant Joseph Ives ventured up the Colorado from its mouth to the western edge of the Grand Canyon in 1857, Young became even more interested in utilizing the river to transport goods and people between Southern California and Salt Lake City (M. Hunter 550–52). In 1861 and 1864, Young directed the settlement St. George, Utah, and the Muddy Mission in southern Nevada as key links along the Mormon-California Trail, sites that would be significant to John Wesley Powell in his Colorado River explorations beginning in 1869 (552–53). Such was

Young's interest in the Colorado that he even enlisted territorial delegate J. F. Kinney to petition Congress for a strip of land on both sides of the Colorado to its mouth so that the church could have an outlet to the sea (M. Smith 39). In 1864 Young also appointed Anson Call to establish a settlement on the Colorado in an effort to explore further the Colorado's viability as a conduit for travel and commerce to support the newly formed Deseret Mercantile Association (553). However, Young's vision of the Colorado never materialized. Call's Landing, or Callville, as the little outpost and its newly constructed warehouse came to be known, was ultimately abandoned because of the river's nonnavigable nature (555). Despite the church's failure to harness the Colorado for its purposes, it benefited from the knowledge acquired by settling within this desert region, knowledge it would share with subsequent visitors to Utah who would throw the Colorado and the Latter-day Saint water culture into the national limelight.

Perhaps the most influential of the earliest observations of the Latter-day Saints' reclamation efforts derive from John Wesley Powell and those participating in his 1869 and 1871 expeditions down the Colorado.[6] During his 1869 journey, Powell carried information that the Latter-day Saint settlers had collected about the Colorado River from the Grand Canyon's western edge to Callville (Stegner, *Beyond the Hundredth Meridian* 100). Upon successfully emerging from the canyon, Powell and his traveling companions gained firsthand experience with Utah's Latter-day Saint settlements and their unique form of water development en route to Salt Lake City (111). As Worster notes, "The Mormons treated them with a ready hospitality that Powell would remember gratefully for years to come" (*River Running West* 194). In preparing to return to the river in 1871 to conduct a more scientific study of the river region and its people, Powell called upon his newly established Latter-day Saint contacts to assist in his safe passage through the region. He met with Brigham Young in Salt Lake City, from whom he secured Jacob Hamblin, likely the most experienced white man in the region, as his guide and interpreter through the canyon country and to the Colorado Plateau's many American Indian tribes (Stegner, *Beyond the Hundredth Meridian* 129).

After leaving the river in 1871 to study the plateau's geography and Indigenous cultures, Powell spent considerable time in Kanab participating in dances and other social events at the local church (138–39). From his experiences traveling throughout Utah and neighboring regions, he penned his *Report on the Lands of the Arid Region: With a More Detailed Account of the Lands of Utah* (1878). Stegner gives significant credit to the Latter-day Saints' influence on Powell's thoughts about reclamation while he traveled through central and southern Utah. From them Stegner writes that Powell "had also got a notion of how salutary co-operation could be as a way of life, how much less wasteful than competition unlimited, how much more susceptible to planning and intelligence, how much less destructive of human and natural resources" (*Beyond the Hundredth Meridian* 227). Powell's own words within the report more simply state his opinion of their efforts: "In Utah Territory, cooperative labor, under ecclesiastical organization, has been very successful" (*Arid Lands* 11).[7] Of Powell's time with the Latter-day Saints, James Wharton concludes in a 1922 treatise of Utah geography and culture, "The importance of this trip through the Mormon settlements, where he saw irrigation methods in active operation upon thousands of acres, cannot be over estimated [*sic*]. Perhaps no one can ever know for certain, but is it not highly probable that it was on this trip he gained that wonderful vision that ultimately led to the organization of the United States Reclamation Service, and the improvement of the lives of hundreds of thousands of American citizens by making their location upon irrigated public lands a reality" (196)? Nearly a century later, Donald Worster would reemphasize the impact the Latter-day Saints' cultivation of the region had on Powell's vision of the arid West. "By the time Powell first encountered them," he writes, "the Mormons had two decades of experience with aridity, and their experience profoundly influenced his thinking and the message he took to Washington" (*River Running* 351).[8]

The Latter-day Saints' reclamation efforts also impressed Powell's close associates. Frederick Dellenbaugh, an artist and topographer in Powell's second expedition, praised the Latter-day Saint settlements and, in particular, Kanab, where he and Powell's party spent the winter of 1872–73. Dellenbaugh admired the charming little village and its water system: "Irrigating ditches ran down each side of every street and from them the water, derived from a creek that came down a canyon back of the town, could be led into any of the lots, each of which was about one quarter of an acre" (*Canyon Voyage* 166). Although this lonely outpost along the Utah-Arizona border had been founded just two years prior to the explorers' visit, the residents had turned Kanab into a latter-day Eden. Dellenbaugh observes, "Fruit trees, shade trees, and vines had been planted and were already beginning to promise near results, while corn, potatoes, etc., gave fine crops.... The entire settlement had a thrifty air, as is the case with the Mormons. Not a grog-shop, or gambling saloon, or dance-hall was to be seen; quite in contrast with the usual disgraceful accompaniments of the ordinary frontier town. A perfectly orderly government existed.... Every one [*sic*] was cordial" (167). Clarence Dutton, a geographer who traveled throughout Utah following Powell's last expedition, also had close contact with the Latter-day Saint settlements. His chapter "Irrigable Lands of the Valley of the Sevier River" within Powell's *Arid Lands Report* corroborates Powell's observations about the church's role in overseeing water development and the cooperative expectation placed on those using water. "This communal arrangement has been attended with great success so far as the development of the water resources are concerned," he writes, "and the system of management has ordinarily been so conducted that the general welfare has been immensely benefitted" (128–29).

The publication and transmission of these observations to government officials and to the popular press led other national figures to praise the church's water use. William Smythe, perhaps the most ardent of the nation's reclamationists, turns to the Latter-day Saint pioneers as a model for sound resource development. In an 1894 article in the *Irrigation Age*, Smythe extols their agriculture in light of the current trends toward monocrops. "Against the folly and error of the one-crop country, east and west," he exclaims, "there is one bright example of success through diversified production which cannot be too often recalled. It is the example of the builders of Utah" ("Republic" 191). Like Powell, Smythe also praised the cooperation that permeated the Latter-day Saint communities. Describing the pattern of Salt Lake City's settlement beginning in the late 1840s, Smythe highlights how every "colonist" became a "small land proprietor, owning a certain amount of irrigated soil from which he could readily produce the necessities of life" ("Mormon Land Tenure" 426).[9] Elwood Mead shared Smythe's views on this point, noting of the Utah experience "how important are organization and public control in the diversion and use of rivers" (*Irrigation* 42). This largely equitable division of land and the communal sharing of water to irrigate ran contrary to the privatization that typified western resource use elsewhere.

One of the more intriguing celebrations of Latter-day Saint water culture comes from President Theodore Roosevelt, who spoke at the Tabernacle on Salt Lake City's Temple Square on May 29, 1903, following the recent passage of the National Irrigation Law. In his opening remarks, Roosevelt celebrates his audience's resolve to redeem a desert and preserve it for future generations: "You took a state which at the outset was called after the desert, and you literally—not figuratively—you literally make the wilderness blossom as the rose.... And you did it because your people entered in to possess the land and to leave it after them to their children and their children's children." Invoking Isaiah's vision as so many others had done before, Roosevelt praises Utahns' radical experiment to make a life in the nation's hinterlands in terms that echo *querencia* and what we might describe today as place-based or bioregional ethics. "You... and your predecessors did not come in to exploit the land and then go somewhere else," he asserts. "You came in... as homemakers, to make homes for yourselves and those who should come after you" (1). Roosevelt contrasts this settlement pattern with those characterized by extractive industries, that quickly seek gain and then move on to the next opportunity. As wise resource conservation was central to Roosevelt's administration, he finds in Utah a model for careful development of the West. In fact, his admiration is so great that he admonishes all Americans, but especially those from the arid West, to "approach the problem of taking care of the physical resources of the country in the spirit with which has made Utah what it is" (2).

Roosevelt elaborates on this stewarding of resources as he turns to the significance of the new irrigation law and its potential for the development of the American West. Once again, he draws upon Utah's pioneer heritage while acknowledging the federal government's indebtedness to its influence on western reclamation:

> And here the Government, to a degree, had to sit at the feet of Gamaliel in the person of Utah; for what you have done and learned was of incalculable benefit to those who engaged in framing and getting through the national irrigation law. For irrigation was first practiced on a large scale in this state. The necessity of the pioneers here led to the development of irrigation to a degree absolutely unknown before upon this continent, and in no respect was the wisdom of the early pioneers made more evident than in the sedulous care they took to provide for small farms carefully tilled by those who lived on and benefitted from them. And hence it comes about that the average amount of land required to support the family in Utah is smaller than in any other part of the United States. (5)[10]

Emerging from Roosevelt's observation and noted in previous accounts of Utah's settlement is the pioneers' adherence to modest living and a careful stewarding of the resources in one's home place. Their particular type of home economics allowed them to increase the per-acre productivity of the land even when living in an arid region.

A final response to Utah's early reclamation history reveals just how influential had been their experience in the Great Basin upon the national impetus for western progress and expansion. Just months after President Roosevelt's Utah visit, the National Irrigation Congress held its eleventh meeting in Ogden, Utah, about forty miles north of the state's capital. To commemorate this event and to celebrate the advances of western reclamation, Virginia McClurg penned the anthem "Prize Ode to Irrigation" with music composed by John J. McClellan.[11] Within this twenty-six-page score, the history of western reclamation unfurls. An opening salute of "O, Desert land!" and paean to both the desert's beauties and its stark realities eventually give way to verses that comprise the "March of the Races," which begins with the Puebloan people's influence in the region, their cultivation of maize, and blessings from the Great Spirit. Following this treatment, she turns to the "Pioneers," whose opening words read, "To the throbbing heart of the fervent earth / Wend the men of the latter day" who "By the dead, salt sea in treacherous guise / Living streams gushed from their urn." Tapping into well-established connections between the Salt Lake Valley and the Holy Land, the verse's chorus reads:

> Fair Cannan smiles in Deseret,
> 'neath Pisgah heights now won.
> Land of the long promised people,
> Green fields, honey bees, and golden sun!
> Land of the long promised people and gold sun!

Considered in light of the biblical prophecy that the wilderness would rejoice, McClurg's ode expresses the nation's belief that the arid West awaited its divinely appointed redemption from a fallen state. She reinforces this point by celebrating the religious refugees seeking a home of their own—like the Children of Israel to whom Isaiah's writings are directed—who would be led by their own prophet to the Promised Land. Such religious rhetoric in connection with Utah's reclamation success even comes from the editor and publisher of the *Irrigation Age*, who enshrines Salt Lake City in the opening pages of the March 1918 issue as the "Mecca of all irrigation activity in the United States" (D. H. Anderson 7).

This early attention to Utah's reclamation efforts has continued well into the twentieth and twenty-first centuries, as contemporary scholars have further entrenched the Latter-day Saints as the model for large-scale irrigation throughout the West. As noted previously, Wallace Stegner's groundbreaking work on John Wesley Powell's Colorado River expeditions not only catapulted Powell into the national limelight as a western hero and prophetic voice for water-resource development but also underscored the influence the Mormon pioneers had on Powell's thinking and the broader development of the West. Stegner writes in *Mormon Country* (1942), "One element in Mormon settlement is invariable: water. Mormonism flowed down the rivers and the irrigable valleys" (38). In doing so, they spread their religion, social structure, and particular relationship to water throughout the Mormon colonies of the Intermountain West. Historian Donald Worster, perhaps not so affectionately, referred to Mormon irrigators as the "Lord's Beavers," who he explains in his classic study of Western reclamation, *Rivers of Empire* (1985), "were the first in the West to propagate assiduously, in deeds as well as words, the gospel of desert conquest sanctioned by God" (74). Worster goes on to trace Mormon settlement of Utah and the Intermountain West, pointing to their industriousness, sense of community, unique structure of authority, and commitment as essential ingredients to their initial success with irrigation, which made them, as he explains, "the first commercially successful irrigators in North America" (75).

More recently, John Fleck, a western water expert, claims in *Water Is for Fighting Over and Other Myths About Water in the West* (2016), "Mormons have been central to the task of measuring, marshaling, and using the West's water in new ways" (44). But perhaps the most direct summation of Mormon influence in reclamation comes from the opening pages of Marc Reisner's *Cadillac Desert* (1986). He acknowledges how the Reclamation Service—later the Bureau of Reclamation—would, in Reisner's view, "be based on Mormon experience, guided by Mormon laws, run largely by Mormons" (2). Not surprisingly then, William Rowley observes in the Bureau of Reclamation's official history, "Industrious enterprise and community cohesiveness inspired by religious commitment in Mormon Utah set an example for how arid lands might be brought under cultivation by irrigation" (56). This sampling spanning nearly a century and a half reiterates the revolutionary and influential nature of the church's efforts to make the desert blossom as the rose upon generations of reclamationists. The Latter-day Saints' success in one of the most arid regions of the continent demonstrated how larger settler communities could not only exist but thrive in the West's arid hinterlands.

The Challenges of Reclamation

Yet such praise tends to overshadow the many challenges and even failures experienced by Latter-day Saints as they established their homes in the region. As Thomas Alexander explains, the narrative of the pioneers' water development tends to "paint an almost idyllic picture of the functioning of the irrigation companies as they transformed the Wasatch Oasis from a desert to a cultivated landscape of towns, farms, and businesses" ("Irrigating" 178). For the pioneers settling in the Wasatch Front's many valleys, they found numerous streams and fertile soil to irrigate (176). However, those settling, either by prophetic call or, later, by choice to the more arid outposts of Deseret, and later Utah Territory, found much more challenging conditions. As Alexander further notes, the hardships pioneers confronted in the territory's more arid regions occurred for a number of reasons, including their failure to "understand the environmental conditions in which they lived," their lack of money to fund projects, and the inability to always secure cooperative labor needed to build the water infrastructure (179). The Latter-day Saint communities also faced difficulties with governments and Indian tribes whose lands they were colonizing (Frehner 185). Out of the 537 Mormon settlements established throughout the late nineteenth century in the North American West, 88 of those communities, or 16.4 percent, ended up failing because of external and environmental factors (Alexander, "Irrigating" 185). Thus, Latter-day Saint irrigation can be described as a "mixed legacy of success and failure" (174).

The historical record is replete with examples of the Latter-day Saints struggling to establish successful communities throughout the latter half of the nineteenth century and even into the first few decades of the twentieth. In 1855 Alfred Billings led a group of forty-one men called by Brigham Young to establish the Elk Mountain Mission in what is today Moab, Utah. Leaving Manti City, they trekked east, struggling as earlier groups of explorers had as they navigated the rugged canyon country of eastern Utah. On May 27, Oliver Huntington, one of the two mission clerks, noted that the party "came to the divide or rim of the Great Basin where the waters began to run towards the Green River and Colorado."[12] Though brief, this observation underscores the group's awareness of the region's geography and the significance of these hydrologic boundaries that cut off the Great Basin from these important rivers to the east. They understood that their success largely depended on their ability to secure water from the Colorado and nearby tributaries to grow crops that would allow them to establish a permanent settlement and thereby extend the church's influence within this rugged part of Utah Territory.

As the group continued in a southeasterly direction, they encountered the Colorado Plateau's difficult topography. On Sunday, June 10, the diarist captures this observation in the mission's official record: "Road extremely sandy, 8 miles and day most exceptionally hot—most of the teams were near giving out when they came to the Canion [*sic*] descent leading to the river—Grand River." Along the Grand (Colorado) River, the record notes farming by the Ute Indians who had cultivated "about 10 acres planted to corn mellons [*sic*], squashes & pumpkins," which were irrigated by "forming dams by which the land was flooded in small quantities—all this was done by hand." The missionaries worked to replicate the irrigation works observed among the Indians and those developed along the Wasatch. However, they quickly discovered how challenging this venture would be in a land of sand and rock. Their first attempt at dam building on June 20 failed, as "being in sandy land it broke away on the night of the 19 and ruined the cite [*sic*]." It was not until they moved upstream later in the day and built their new reservoir on an existing beaver dam that they diverted consistent water to their newly plowed fields. Yet even with seeds in the ground, the challenges continued. On June 21, Huntington laments how "considerable grain and potatoes were put in the ground, but quicksand in the main ditch made considerable trouble." He also notes how the "river had fallen rapidly," a sign that the spring runoff from the Rocky Mountains had ended (*Elk Mountain Mission*). The preoccupation with water fills the journal, no doubt because its presence or lack thereof meant the difference between survival and defeat, or possibly even death. However, the mission did not endure long enough to see if their rudimentary irrigation works would last. Despite initial success in converting some of the local Indians, violent skirmishes in which both Indians and missionaries were killed led President Billings and the remaining missionaries to abandon their newly erected fort and return home.

Latter-day Saint villages along Arizona's Little Colorado River further dramatize the difficulties of reclaiming the high desert. Beginning in 1876 and continuing through the 1880s, the church established a number of villages along the Little Colorado as part of a greater plan to extend the church's influence into Mexico (Peterson 6).[13] These included places like Brigham City, St. Joseph, Woodruff, and St. Johns. William Abruzzi's extensive study of the development of the Little Colorado reveals the significant difficulties these communities faced. As with other colonies, those along the Little Colorado depended on impounding water behind dams to irrigate crops. Yet without adequate funds and labor, the settlers built small earthen dams composed of rock, dirt, and brush that easily fell victim to washing out through seasonal runoff or cloudbursts. As Abruzzi notes, the dams on the lower reaches of the Little Colorado around St. Johns and Woodruff were particularly susceptible to failure. In fact, between 1876 and 1923, fourteen dams in St. Johns collapsed (123). In 1915 the people of St. Johns contributed their own funds to those offered by a Colorado construction company and church headquarters in Salt Lake City to build a new dam. Despite the hundreds of thousands of dollars it cost to build the structure, it collapsed, with the subsequent flood killing eight people (132–33). Such regular tragedies tested the resolve of the settlers, some of whom chose to leave rather than continue to fight the desert (128). Yet as Charles Peterson notes, for those settling along the Little Colorado, "The thought that a life close to the earth was evocative of the highest human virtues was also important in Latter-day Saint thinking. Joining the idea with a firm conviction that hard work and frugality contributed to spiritual growth, Mormons were confident that the unproductive lands of their desert kingdom could scarcely have been better calculated to foster virtue among the Saints" (156). They believed their mission was divinely appointed. On the land they would prove their faithfulness by making the desert bloom and, in so doing, feel God's approval and become worthy of their sacrifices made to establish homes in such a forlorn land.

While Latter-day Saints faced numerous external challenges that continually tested the settlers' resolve and faith, internal conflicts also complicated the typical narrative of Latter-day Saint irrigation that so many observers have praised. While cooperation and community are virtues typically synonymous with the Latter-day Saint colonization of the Intermountain West, such traits were continually challenged by the lack of water. John Bennion's 2019 novel, *Ezekiel's Third Wife*, captures these sentiments and juxtaposes them against the dogged determinism within these rural outposts to maintain the purity of the community. As drought grips the fictitious town of Rockwood, set on the fringes of Utah's West Desert in the 1890s and within the waning years of the church's practice of polygamy, the close-knit connections between neighbors begin to unravel as Sophia, Ezekiel Wainright's second wife, is found dead in the irrigation ditch. At the center of the investigation of her murder is Rachel O'Brien Rockwood Wainright Harker, Ezekiel's third wife. Unbeknownst to Ezekiel and the rest of the community, Rachel has secretly married Matthew Harker, a man not of her faith, with whom she plans to run away to start a new life outside the constraints she finds within plural marriage and the challenges of farming this desert outpost. On the night when Rachel is to meet Harker in the barn for a late-night rendezvous, she discovers Sophia's body in the ditch and begins a frantic search for both the murderer and her lover, whose tracks lead past the ditch and into the desert. Fearing that her father-in-law, J. D. Rockwood, town founder and expert tracker, will hunt down and kill Harker, whose tracks Rockwood believes are the murderer's, Rachel takes it upon herself to solve the mystery of Sophia's death to keep her illicit relationship under wraps. As the novel's plot centers around the irrigation ditch of this small Latter-day Saint community, the complications associated with communal desert living emerge.

Well into the plot, Rachel provides a glimpse into her early life that brought her, her brother, and her mother from Nevada's mining fields and an abusive father and husband to Rockwood, where she would eventually become part of J. D.'s family. Her new home and its surroundings typify the small Latter-day Saint communities throughout Utah and neighboring states while contrasting with the extractive industries enticing many to go west during this period: "We lived in a green valley, never as green as Ireland, but green from irrigated fields, orchards, even a flower garden or two. Tall Lombardy poplars had been planted as a windbreak and for beauty.... We walked along ditches full of water. The earth wasn't pitted with mines, was not marred with stone and dirt rubble" (*Ezekiel's* 71).[14] Bennion portrays Rachel's new surroundings in bucolic idealism. Yet as drought sets in, she finds life in Rockwood becoming increasingly difficult. She explains, "As June moved toward July, the weather had grown even hotter and drier, the white alkali desert shimmering in the distance, even the willows wilting, as if the whole town would explode into flame.

Violent words and acts burst out on the streets and in church. People glared at each other as they watched the shrinking stream.... Against my own better judgment, I found myself just as angry over water as everyone else. The lower half of my fields dried to mere dust." Struggling against these conditions, Rachel concludes, "The lack of water was breaking apart our Zion community" (12).

Tensions only increase in the town as the water master, Brother Swensen, along with the bishop, decide on a new rotation of water deliveries to occur every twelve hours. Committed to equity and cooperation, Brother Swensen tells his neighbors and fellow worshippers, "I'd rather the Saints die and be committed to their graves than to watch them steal from one another" (12). While such ends seem extreme, Rachel notes that the inability to cultivate crops and provide for one's family can make people into something they are typically not. Though she believes that "Rockwood was a town of good people," she also sees that privations bring out the worst in them: "Only lust for water was strong enough to make a good person kill" (107). Through Rachel's keen observation and quick thinking, she eventually discovers the murderer: none other than the water master, Brother Swensen, whose duty to his position and his community overcomes his sense of reason and decency. In the end, Rachel finally reunites with her husband Matthew, with whom she flees her Latter-day Saint community and family in search of new opportunities and freedom.

While Bennion's fictional account represents water conflict and the perils of community in the extreme, the tensions experienced within these desert outposts are not without precedent. Bennion's inspiration for the novel draws from the journals of his great-grandfather Israel Bennion and the struggles the town of Vernon, Utah, experienced at the turn of the century as the community vacillated between "two ideological positions as Utah shifted from communal to capitalistic management of water" through prior-appropriation doctrine. Captured in Israel's 1893–96 journals, which cover the years leading up to Utah statehood in 1896, the conflict stems from a host of factors, including the transition between water-governance systems but also the growth of a non-Latter-day Saint population wary of church control of natural resources and the state's increased economic activity (Bennion, "Water Law" 289). As a counselor in the town bishopric, the local ecclesiastical unit that typically oversaw water allocations in the Latter-day Saint communities, and secretary of the Vernon Irrigation Company, Israel found himself clashing with Erick John Pehrson, the second counselor in the bishopric and vice president of the irrigation company. While Bennion and the bishop, John Sharp, supported the old system of church-governed water deliveries, Pehrson embraced privatized water (295). As each group moved to develop the valley's streams according to their respective beliefs, conflict ensued. When Sharp and Bennion built a new ditch to aid the community, Pehrson saw his right to the water under prior appropriation under assault (297). Eventually, the town fractured over these two approaches to the point that the opposing factions would not gather for the annual Fourth of July festivities (302). In Israel's mind, years of disagreement, reluctance by town members to work on ditches, and even destruction of property stem from "a too great love for the things of this world, and too little appreciation of the 'unspeakable things of the Kingdom.' A dollar is allowed precedence of the love and fellowship of our brother" (299, 304). As with most Latter-day Saints during this period, Israel sees the health of the community as more important than the prosperity of the individual, yet his efforts to unify Vernon fail as the town, like the rest of Utah, ushered in a new age of water distribution that continues to this day.

Utah Reclamation Today

The colonization of Indian lands, irrigation of terrain marked by sparse and unpredictable moisture, and the internal conflicts between community members reflect some of the many issues confronting Latter-day Saints as they made the Intermountain West their new home. Notwithstanding these many challenges and the failure of some colonies and missions along the Colorado River and more arid parts of the West, Utah's present landscape shows that in many cases, the Latter-day Saints successfully fulfilled Isaiah's vision of a fruitful desert. In fact, we might say today that they were too successful in their attempts to redeem the desert. Where Utah was once the envy of the western states in terms of its irrigation success, it has become a model of excess and waste in recent years. While Utah is one of the nation's driest states, it also carries the dubious distinction of being America's "thirstiest," as it uses the most water provided by public suppliers (from utilities versus private wells) per capita of any state in the nation—240 gallons a day per person according to the Utah Department of Water Resources (Weiser; Utah's Regional ES-1). This stems from Utah's widespread use of "secondary water," the untreated irrigation water once used to water crops but now used in urban areas to water lawns (Weiser). Because secondary water is not metered, home owners end up using far more water than needed and do so largely because the cost of using this water, in addition to the indoor treated water, is minimal (Weiser). While a 2016 study of Utah's water use recommended secondary metering to reduce waste, not until 2019 was legislation introduced to mandate metering on all existing water use.[15] However, this bill did not pass, although amendments now require meters for all new construction using pressurized secondary water (O'Donoghue).

While such changes are a step in the right direction, there needs to be more immediate action to curb this overuse. In response to the many headlines about Utah's water-wasting habits, Mark Milligan concludes in an article for the Utah Geology Survey, "To sustain such growth Utah will soon need to make big decisions about conservation measures, water management, and the potential development of costly new supply projects" (12). Yet other water experts speak to the need to reassess the state's water situation in more urgent terms. Eric Ewert, a professor of geology at Weber State University, has outlined many of the causes for Utah's impending water woes. These include the state's exploding population and diminishing available resources, as climate change is projected to make significant impacts on the West's rivers and lakes (14–16). After describing the various scenarios that will further imperil the state's water and the possible solutions to curb disaster, he concludes, "We must become more water wise *right now* or else face a various precarious future" (26; emphasis added). Rampant population growth, unpredictable runoff, and major projects like the Lake Powell pipeline to address Washington County's housing boom speak to the need for immediate action to ensure that the kind of life that has made Utah so desirable in the past is possible for future generations. Although the state's Division of Water Resources has made substantial efforts to better inform the public about Utah's water resources through a "Slow the Flow" campaign and an app aimed at helping home owners know how much to water their lawns at certain times of the year, much more is needed to encourage greater conservation ("About the Weekly Lawn").

This recent attention given to Utah's high water use raises important questions about the motivation behind such water waste. Yes, the obvious culprit is the system that continues to allow for secondary metering. But what about more fundamental reasons? What caused Utahns to once be the bell of the ball as far as water use was concerned to now being what many perceive as leaders of profligate waste? Is it because the nation's sentiments toward water use have changed while Utah's has not? Or is it because the pioneers' early reclamation efforts encouraged waste and this legacy has just persisted? Among the many factors that may have led to this transition from careful stewardship to a more disengaged relationship with water is what Thomas Alexander describes as a forgetting, a "gradual loss of the collective memory of a salutary environmental theology" ("Lost Memory" 48). Occurring at the end of the nineteenth century, this transition to "a memory of progress under secular entrepreneurship...played a central role in the loss of environmental consciousness in the Mormon community." Alexander traces this shift to the improved economic conditions the Latter-day Saints experienced, particularly after the completion of the transcontinental railroad in 1869, which opened Utah to a much larger extent to the markets in California and back east (49).

Notwithstanding the gradual turn away from this salutary environmental ethic that informed the early pioneers' experience, various scholars and church leaders throughout the twentieth century have reaffirmed this knowledge and practice. Alexander describes these efforts as "rememberings," and they are evident in the teachings of a few church leaders at the beginning of the twentieth century and during the 1970s–1990s, the latter period demonstrating a significant accumulation of scholarly work about the connections between Latter-day Saint theology and the environment to coincide with the emerging fields of environmental history and ecocriticism (49–50).[16] This act of remembering has only accelerated in the twenty-first century, and particularly within the past decade, as the church institutionally and as members individually highlight the intersections between the doctrine of the Church of Jesus Christ of Latter-day Saints and environmentalism. We might consider this period a third remembering or awakening evident in a number of recent publications, the advent of the advocacy groups LDS Earth Stewardship and Mormon Environmental Stewardship Alliance, and more public statements by the church on its website.[17] As part of this reinvigoration of the church's teachings and their relevance to current environmental issues, I continue this act of remembering through an examination of the writings of John A. Widtsoe and the principles of stewardship and provident living taught within the church. Together this discursive ecology of water literacy shaped by poetic, rhetorical, and scientific registers reframes our understanding of western reclamation by positing alternate water realities to those that have more recently come to reflect Utah's and the Colorado River Basin's water use.

"Reclamation," as understood in the context of making the desert bloom, is defined by the *Oxford English Dictionary* as the "conversion of wasteland . . . into land fit for use, e.g. for cultivation or construction." This particular definition, first given in 1810 to describe the draining of excess water in Ireland, has become the primary means of understanding western reclamation. Other definitions, including one from 1629, also give insight to this term and its physical manifestations: "The action, fact, or process of reforming a person morally and spiritually; the saving of a person *from* a way of life considered immoral, dissolute, or otherwise undesirable; reformation, moral improvement; redemption, reformation." Since its inception, western reclamation has been the process of redeeming a lost and fallen land, making it productive and fruitful. But such definitions of reclamation, when considered in light of today's water challenges, urge us to think differently about the redemption of western lands. In fact, one of the more recent definitions of "reclamation" is just this: "[a] reassertion of a relationship or connection with something; a re-evaluation of a term, concept, etc., in a more positive or suitable way" ("Reclamation"). Thus, what might reclaiming reclamation within the context of our current water challenges suggest? If we return to the 1629 definition, we might reimagine Utah's water use through the lens of an ethical or moral reformation for the purpose of improvement—in this case to ensure that there is sufficient water for our present needs and for future generations. As with the early pioneers who settled in Utah and throughout Deseret, they recognized the importance of cooperation and had a keen awareness of the geographical limits of the places in which they resided. Faith in a better future, sacrifice through cooperation for the greater good, seeing outside our immediate purview, and creating a greater ethic of place are hallmarks of this new western reclamation.

Such goals are certainly aspirational, and one might wonder how relevant Utah's unique culture and history is to the rest of the watershed and other regions facing water shortages. In fact, as some historians have suggested, Utah's pioneering efforts in irrigation were significant for the state's development but did not really translate well elsewhere. As Leonard Arrington and Dean May explain, "The Mormon example persuaded other settlers that extensive irrigated agriculture was possible, but when settlers from California, Colorado, and other states and territories looked to Utah for detailed instructions on how to build and maintain an irrigation system they were most commonly thrown back upon their own experience and resources" (10–11). While this observation may be true to the extent that the organizational structure of the church and its influence on managing water in Utah's early years, as well as the devotion to a cause, could not be replicated equally elsewhere, this argument fails to consider John A. Widtsoe's influence on the development of irrigated agriculture on the arid lands and the value his thoughts on water offer today.

John A. Widtsoe, a Man of Science and Faith

As a world-renowned scientist, a pioneer in irrigation technology, a tireless advocate for Utah's higher-education system, and an influential leader in the Church of Jesus Christ of Latter-day Saints in the first half of the twentieth century, Widtsoe offers us important lessons today as we consider the relationship between science and art, science and religion, and humans and the earth. True, Widtsoe relies heavily upon the conservation discourse of the late nineteenth and early twentieth centuries to which men like President Roosevelt appealed. Not surprisingly, Worster describes Widtsoe as "another passionate believer in the religion of irrigation" (*Wealth* 120). However, this summation is far too simplistic a characterization. While Widtsoe wholeheartedly commits himself to scientific observation and study to advance the reclamation of the West's arid lands for agriculture, he also articulates a relationship with nature that science cannot measure. Through poetic discourse that seems to stand in stark contrast with his utilitarian views of resource development, Widtsoe reflects on the beauty and power of the Colorado River and its canyons to enliven the soul and express his innermost gratitude for life and the creation of the natural world. Writing within a tradition reflective of Alexander von Humboldt, William Whewell, and Henry David Thoreau as discussed in chapter 2, but more pointedly informed by the doctrine of the Church of Jesus Christ of Latter-day Saints, Widtsoe reveals his dedication to the betterment of the human race through a holistic approach to truth. His writings about irrigation and the Colorado that span various discourses construct a new truth about water, or what we might call, in Beale's words, reality, as he embraces the contribution that science, art, and religion make to understanding the human relationship to the natural world.

Born on January 31, 1872, in the small, isolated town of Dalöe, Norway, Widtsoe grew up in a family who prized education, his father being a schoolteacher. After his father's death when Widtsoe was six years old, the family moved to Trondheim, Norway's capital at that time. Devout Protestants, the family read regularly from the family Bible, but Widtsoe's education was far more than just religious. He spent many happy hours reading from his father's library, recalling with delight *The Merchant of Venice*. As he explains of his early years and his development as a reader, "Fairy tales, and the ancient sagas, developed my imagination; and indeed fixed upon me a taste for imaginative literature, which I have valued throughout my life" (*In a Sunlit Land* 5). When Widtsoe was nine years old, his mother joined the Church of Jesus Christ of Latter-day Saints; Widtsoe would later convert. With his mother and brother, Osborne, he eventually immigrated to Logan, Utah, where he arrived on November 17, 1883 (8).

Widtsoe's love of learning eventually led him to Brigham Young College in Logan, Utah, and then to Harvard University, where he studied chemistry, although the choice came after some struggle. He recalls, "Mathematics and English battled for recognition but my Scandinavian common sense determined my decision" (31). Following his graduation from Harvard, Widtsoe took a position at the Utah Agricultural College in Logan, a place where he would pioneer irrigation science and further his development as a consummate educator. In 1898 he and Leah Dunford, whom he had recently married, traveled to Germany's University of Göttingen, where Widtsoe pursued graduate studies in chemistry. After returning to Utah in 1900, Widtsoe became the director of the Utah State Agricultural Experiment Station, also in Logan. Widtsoe's success there led to new positions as presidents of both the Utah State Agricultural College (USAC) and the University of Utah, as well as the president of the International Dry-Farming Congress. In 1921 he was called as a member of the Quorum of the Twelve Apostles, one of the church's highest governing bodies. This new role took Widtsoe to various parts of the globe directing the church's affairs. He served in this position until his death on November 29, 1952.

Although an immigrant from well-watered Scandinavia, Widtsoe developed a "life-long love of the desert" and emerged as a world expert in irrigation science, as well as the environmental constraints relating to Colorado River development (*In a Sunlit Land* 50). Utah had inspired many scientists and irrigators in Powell's day, and a generation later, the state continued to draw officials from all over the world to see the work that Widtsoe and his colleagues performed at the experiment station (119). Of the experiments pursued there, he writes, "It seemed only proper that they should have been undertaken in Utah, the first state to settle in the heart of the desert" (76). From this and subsequent work, Widtsoe wrote such acclaimed works as *Dry Farming: A System of Agriculture for Countries Under Low Rainfall* (1910), *The Principles of Irrigation Practice* (1914), and *Success on Irrigation Projects* (1928), which brought great notoriety to the station and to Widtsoe, specifically. Of Widtsoe's work on dry farming, Dr. Eugene W. Hilgard, an agricultural scientist of national prominence, remarked, "It should be in the hands of every farmer from the Plains country to the Pacific Coast" (qtd. in Parrish 193). A *Deseret Evening News* article concluded of Widtsoe's contributions at the USAC and the experiment station, "This work established him as one of America's foremost scientists and placed the Utah Experiment station among the first few in the world" (qtd. in Parrish 163).

While Widtsoe eagerly engaged those coming to Utah, he enjoyed traveling throughout the West and even internationally to help farmers more efficiently use their water. Central to Widtsoe's work was identifying how much water to apply to different crops to produce successful yields. In response to an 1897 trip throughout Utah to better understand the state's soils, he recalls, "The endless sage-covered deserts lay before us. Sage, greasewood, rabbit brush and other native plants, we learned, were safe indicators of varying soil quantity. I was fired with a desire to help cover these millions of acres with crops valuable to man. The dream that such a conquest is possible remains with me" (*In a Sunlit Land* 50). While such rhetoric goes against the current of today's environmental sensibilities, it represents a particular moment in time when the nation looked to develop the West's resources to advance the country and region into a new era of prosperity. Utah was still a relative backwater in 1897, having become a state only the year before, and many within the state still lived in primitive conditions. Not surprisingly, Widtsoe, like most Utahns of that period, desired to tame the harsh conditions of the desert to create more comfortable and predictable living.

Despite Widtsoe's rhetoric that echoes other contemporary proclamations about redeeming the desert, he was also a realist and understood—as had Powell decades before—that there was only a finite amount of water available in the West. Speaking of Utah, Widtsoe explains, "Only a small portion of the desert could be irrigated, even under the most favorable conditions, with the water flowing from the mountains" (74). This realization is powerful, particularly considering this period when the federal government was ramping up efforts to reclaim as much of the arid West as possible. Also important is Widtsoe's understanding that the application of water needs to be done responsibly, according to true principles that have been well tested through extensive experimentation and observation. In summarizing his research at the station, he notes, "After a certain quantity of water is applied to a crop, further additions do not increase the yield or increase it very lightly" (117).

Widtsoe reinforces responsible water use in a pamphlet he drafted for the United States Food Administration for Utah. Entitled *Don't Waste Water*, the publication expresses Widtsoe's disdain for the waste he observes "from Richmond in the north to St. George in the south." The document reflects Widtsoe's concern that such waste fails to follow sound scientific principles—a common refrain in his writings that highlights how opinions should matter little when confronted with incontrovertible fact. He writes, "The farmer often thinks that the more water is used, the larger the yield. This is not so." At the same time, however, his complaint has just as much to do with ethical behavior as it does with science. He observes that such waste represents a failure to care for one another, to ensure the success and prosperity of the community. "Careless of our own need and the cry for the world for bread," he argues, "we stand by and watch in apathetic silence the waste of life-giving water; or worse still, unmindful of our brother's right, or ignorant of the harm we do, we sanction the waste of water." Widtsoe recognizes the prior-appropriation doctrine as he speaks of "our brother's right" and its related demand that one's use causes no harm to other irrigators. However, we can read Widtsoe's admonition as a refrain equally steeped in his concern for the community, emanating from Utah's earlier water history and the theological teachings of his faith. Acutely aware of his predominantly Latter-day Saint audience, Widtsoe outlines the harm caused by overirrigation to the soils and to other farmers, particularly along the benches of Utah's valleys, and asks, "Is the man on the bench his brother's keeper?" In referencing Cain's response to Jehovah in the Old Testament, Widtsoe advocates an ethic that requires one's commitment to the community's well-being rather than individual gain.[18] To further emphasize this point, he rebukes the "greedy water hog, who begs, takes and misuses water, simply to keep it from his neighbor—to maintain his claim upon it, as he says."

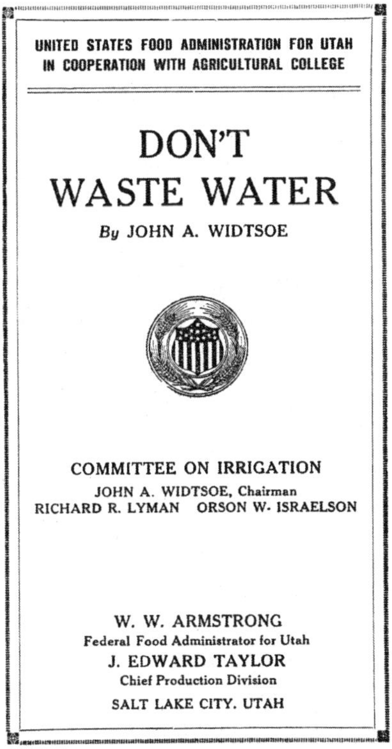
UNITED STATES FOOD ADMINISTRATION FOR UTAH
IN COOPERATION WITH AGRICULTURAL COLLEGE

DON'T
WASTE WATER

By JOHN A. WIDTSOE

COMMITTEE ON IRRIGATION
JOHN A. WIDTSOE, Chairman
RICHARD R. LYMAN ORSON W. ISRAELSON

W. W. ARMSTRONG
Federal Food Administrator for Utah
J. EDWARD TAYLOR
Chief Production Division
SALT LAKE CITY. UTAH

Don't Waste Water. John A. Widtsoe. Box 3, Folder 4, ACCN 1562 John Andreas Widtsoe Papers, Special Collections and Archives, University of Utah, Salt Lake City.

Also apparent in this tract is Widtsoe's place consciousness and his attempts to instill within Utahns a new narrative of reinhabitation. Observing how waste is often created by unmaintained water-delivery systems, he writes, "Let a leaky canal be a disgrace to those who own it. Stop that waste of water, if you love your country!" While Widtsoe's reference to country might seem a more patriotic statement, his focus on Utah farmers suggests that this reference invokes one's regional affiliation as opposed to national devotion. Here one sees Widtsoe's bioregionalism emerge. His preoccupation about water use begins as a critique of the poor application of tried-and-true scientific principles and then transitions to an invitation to embrace a higher law that eschews individualism for an ethic of community care. Deeply concerned that water is available to benefit the greatest number of people possible, Widtsoe invites his readers "to use water as to make possible a large increase of irrigated farms" (*Don't Waste Water*). The irony here is that by exercising restraint and curtailing one's excess, the land will be more productive and able to feed more people through this approach to water savings.

Notwithstanding the success Widtsoe found throughout his experiments, his fellow Utahns were not always ready to accept the new science he promoted, especially when they discovered that he himself was not a farmer. Once when meeting with a group of farmers to present new information on how to more efficiently water crops, one of the men looked at Widtsoe's "soft hands" and explained how he rued the day when a man with no farming experience would tell him how to manage his fields. However, as Widtsoe explains, "It did not take long to conquer the people of Utah. They needed only to be convinced that we came as bearers of truth and that 'book-farmers' so-called had much to teach the pioneer who had to learn only through hard experience. Then, the old prejudices vanished" (*In a Sunlit Land* 79). Voiced more than a century ago, such words carry significant weight today, as a substantial portion of Utahns ignore or reject science-based evidence for such issues as climate change.[19] In addition, as we reflect on Utah's history of excess water use, Widtsoe's words suggest that if the most ardent defenders of irrigation could change then, Utah's citizens can change now, as they possess the facts about current water trends and their impacts on the state's future.

Amid this confluence of science, community, the greater good, and religion that emerges from Widtsoe's beliefs about water use are his more poetic expressions of nature, evident in Widtsoe's journal from a 1922 Colorado River trip. This journey, conducted as part of a federal investigation of potential dam sites in the Glen Canyon area and in preparation for the Colorado River Compact deliberations that culminated two months after the trip, reveals another side to Widtsoe. His musings capture his deep reverence for nature to complicate the science-religion and science-poetry tension that has long defined which epistemologies should shape resource decision making. With R. E. Caldwell, the Utah state engineer, Widtsoe joined influential men including representatives from the region's major energy and transportation sectors, as well as Arthur Powell Davis, John Wesley Powell's nephew and director of the Reclamation Service; E. C. LaRue, discussed in this book's introduction; and Clarence C. Stetson, secretary to Herbert Hoover, the Colorado Compact commissioner. Departing central Utah for the river, Widtsoe remarks on September 6, "There is so much magnificent scenery that in spite of our best desires, we are not appreciating it as we should.

Marvelous how man becomes callous!" (*A Journal* 202). His concession suggests a quality all too easily overlooked or forgotten during the hectic nature of everyday life.

Yet the magnificence of his surroundings invites a different kind of observation and perception from his usual scientific work. A day later, as the party arrives at the river, he exclaims, "God be thanked for the earth and its beauty; for life; for hope" (206). Widtsoe's view of the river and canyon as God's handiwork permeates the journal, as do descriptions of the creations steeped in the language of a poet. When the travelers finally push onto the river on Friday, September 8, Widtsoe gushes: "The scenery is sublime. Great cliffs of majestic proportions and vivid red, and of infinite variety. Immense arches, auditoriums, stadiums, amphitheaters, temples, palaces, nature's writing on the walls. There are hanging gardens, groves and glens, apparently blind alleys that turn everywhere as you come nearer. Occasionally mighty rocks tower out of the river. There is an overpowering quality in the feeling induced by the scenes along the River. One of the most notable days of my life!" (207). This last sentence underscores the trip's influence on Widtsoe's understanding of the river canyon from a spiritual, soul-enriching point of view instead of one fixated on their geologic processes. Nowhere in this passage appears a description of the canyon in objective, scientific terms. Rather, Widtsoe couches his observations within the Romantic sublime, thereby acknowledging the tremendous works of nature that bring the viewer to a recognition of the divine. Floating downriver, he endows nature with agency as he witnesses its handiwork or "writing" etched upon the canyon's infinite variety of colors and shapes playing out before his eyes.

Such exhilaration and sentiment continue during the next week. He observes "high walls on both sides of us; then breaking into lower walls, especially on east side. The water smooth as a lake. Blue heron on the shores; green patches in crevices of the high walls (about 1200 ft. high)." While a brief notation of elevation attempts to quantify the height of the surrounding cliffs, the cataloging of beauty that fills his entry overshadows this point: "Many little side canyons, great arches, caves, amphitheaters; cover; turns, bends; pinnacles; towers, domes; galleries; paths; reflections of wondrous delicacy, broken and softened by the ripples. As the sun rises, the light reflected from the ripples makes the water a golden yellow; where the shadows still lie, the light from the rock walls make the water a lustrous red and gold. It is altogether wonderful" (214). This wonder fills the rest of the journal as each day reveals another splendor of God's creations. Widtsoe describes Rainbow Bridge as "a perfect bit of the Master's work," and later, as the group arrives at Lees Ferry, the morning "is one of brilliant immensity. This is the Lord's Day, and we worship Him this day through his works (219, 227). A sunrise inspires him to ask, "Shall I ever see anything so charmingly appealing again? The clear sky, the firmament of stars and the full moon, have made our nights wondrous in their beauty" (221).

As the party floats peacefully down the concluding miles of their voyage, Widtsoe reflects on the impact the trip has had on him: "Looking back this has been one of the great trips of my life. The scenery has been so vast, varied and awe inspiring. The solitude has been so complete. The beautiful courses of canyons and glens have been so full of beauty. The slow flowing river has been so thought promoting. The clear sky and the starlit and moonlit nights have brought us so near to the meaning of things 'as they are.' The whole result has been one of continuous, wholesome elevating and especially soul stimulating enjoyment" (223). This sensory feast for the soul fills Widtsoe's spirit, evoking emotions that contrast starkly with his other water-focused works filled with calculations and figures about how best to use Utah's water. Within his private musings, he expresses gratitude and wonder for the world his God created and for this experience that brings Widtsoe to a clearer sense of reality or truth expressed as "the meaning of things 'as they are.'" For Widtsoe, as for Childs, scientific discourse fails to capture the scenes and feelings of this unforgettable trip. This broader perspective concludes Widtsoe's journal as he agrees with another renowned scientist, John Wesley Powell, "that it is useless to describe with words or even with pictures the wonders, of surpassing magnitude and beauty, that fill the country through which we have passed on this trip" (229). In the end, it is the language of feeling, of sentiment, of the poet that can best approximate the marvel and wonder of the Colorado River and its canyons.

Stewardship and Provident Living

These selections of Widtsoe's extensive writings reflect various impulses. As a leading expert in irrigation science, he assisted Utah, the nation, and other countries as they confronted the constraints aridity imposes on agriculture. Sound practices led by careful studies and proven results dictated his approach. But as his instruction on water waste demonstrates, water is a precious resource that, while bound by legal principles supporting individual consumption, must be used to benefit the larger community to ensure that each member prospers. His writings also complicate the idea that nature must be tamed. Although he routinely expresses his opinion that man is destined to control the earth—a perspective developed more fully in this section—he also waxes eloquent as he lingers on the supernal beauty of the Colorado River and the extraordinary assemblage of plants, animals, rocks, and water that fill his soul to rejoice in the great beauty of the world and power that nature has in uplifting humankind.

Widtsoe's views of nature may seem at odds, but we may better understand them when read through the principles of stewardship and provident living as taught within the Church of Jesus Christ of Latter-day Saints. Stewardship, for many Christians, implies humans' responsibility for the divine creation of the earth—a position that reframes the Genesis account of Creation wherein humans receive dominion over the earth. Stewardship teaches that while humans hold a significant position within the Creation, they carry a monumental responsibility in caring for the planet they have inherited. Of this relationship between humans and the earth, Elder Marcus Nash, a leader in the church's Quorum of the Seventy, teaches, "[God] intends man's dominion to be a righteous dominion, meaning one that is guided, curbed, and enlightened by the doctrine of His gospel—a gospel defined by God's love for us and our love for Him and his works. The unbridled, voracious consumer is not consistent with God's plan of happiness, which calls for humility, gratitude, and mutual respect" (268).[20] Thus, this "dominion," rather than a heedless mandate to control and conquer the earth, requires restraint in one's daily desire for material possessions, a heightened appreciation for the natural world, and the recognition that everything within the earth—whether human or not—occupies a divine position in the Creation. That is, they do not merely exist through happenstance, but exist because they occupy a central purpose in God's designs for humanity. In the Pearl of Great Price, a sacred text within the church, God teaches Moses that "out of the ground made I, the Lord God, to grow every tree, naturally, that is pleasant to the sight of man; and man could behold it. And it became also a living soul" (Moses 3:9). As this scriptural passage suggests, trees not only bear aesthetic qualities to bring joy to humankind but also have souls. The text also emphasizes this with all animal life, as all species are "living souls" (Moses 3:19). Elder Nash elaborates on these verses, explaining, "Since both plant and animal life are living souls, they are capable of experiencing happiness as they fulfill the measure of their creation." He then quotes Joseph F. Smith, the sixth president of the church, who teaches, "It was intended that all creatures should be happy in their several elements" (267). From this commentary, the church's doctrine makes clear that human's relationship to the earth is one that must allow for all of Creation to flourish within its own sphere—that they are of value in and of themselves.

However, this position does not imply that these creations are not to be used. Again, the scriptures clarify where Latter-day Saints stand on this relationship. The Doctrine and Covenants, another sacred text used in conjunction with the Bible, Book of Mormon, and Pearl of Great Price, captures Jesus Christ's words for humanity:

> The fulness of the earth is yours, the beasts of the field and the fowls of the air, and that which climbeth upon the trees and walketh upon the earth; Yea, and the herb, and the good things which come of the earth, whether for food or for raiment, or for houses, or for barns, or for orchards, or for gardens, or for vineyards; Yea, all things which come of the earth, in the season thereof, are made for the benefit and the use of man, both to please the eye and to gladden the heart; Yea, for food and for raiment, for taste and for smell, to strengthen the body and to enliven the soul. And it pleaseth God that he hath given all these things unto man; for unto this end were they made to be used, with judgment, not to excess, neither by extortion. (59:16–20)

In these oft-quoted passages among Latter-day Saints, the Lord teaches that all Creation is for humans' temporal and spiritual use. That is, the earth's unsurpassed beauty, its infinite variety, and its mere presence should uplift humankind and evoke sentiments of gratitude and joy similar to those feelings expressed by Widtsoe in his journey down the Colorado. However, the Lord provides an essential caveat here: The earth's resources are to be used wisely and in moderation. To use these creations to sustain and enrich the human family is expected; to pursue wealth and the accumulation of things for the sole purpose of one's own pleasure is contrary to divine purpose. Widtsoe's admonitions against waste are clearly evident in this theology.

Yoked with stewardship, provident living has long been taught in the church to prepare its members for times of difficulty and aptly describes the approach early Latter-day Saint water development took to providing for farms and cities, as well as Widtsoe's pioneering work in irrigation science. Provident living encompasses self-reliance and preparation, the idea that people should live within their means and cultivate the necessary skills to take care of themselves and others—qualities President Roosevelt praised in his address to Utahns. It defines the church's approach to welfare, its plan for caring for the poor and needy. As Joseph Smith, the faith's founder, is quoted on the church's welfare website, "A man filled with the love of God, is not content with blessing his family alone, but ranges through the whole world, anxious to bless the whole human race" ("Proving in the Lord's Way"). More recently, current church president Russell M. Nelson teaches how one's stewardship of the earth assumes care for others: "We should care for the earth, be wise stewards over it, and preserve it for future generations. And we are to love and care for one another" ("The Creation"). Sharon Eubank, first counselor in the church's General Relief Society Presidency, reiterates this unique relationship between caring for the earth and others in an address to LDS Earth Stewardship.[21] When asked whether we should be "caring for the poor versus caring for the Earth," she responds, "Are they not linked so inextricably that we can't do one without caring for the other?" ("That We May Be One"). Articulating an ethic of care both for the earth and for others in what is essentially an argument in favor of environmental justice, Eubank underscores the inextricable link between stewardship and provident living. Thus, while Latter-day Saint scripture teaches that "the earth is full, and there is enough and to spare," this does not condone heedless spending and wanton consumption of the earth's resources (Doctrine and Covenants 104:17). Instead, when understood through provident living, it is clear that the earth can provide all the necessary resources, such as clean water and food, only *if* the world's inhabitants choose to alter their actions. Behavior to the contrary solicits a very firm rebuke: "If any man shall take of the abundance which I have made, and impart not his portion, according to the law of the gospel, unto the poor and the needy, he shall, with the wicked, lift up his eyes in hell, being in torment" (104:18). Therefore, provident living reflects one's willingness to eschew greed and materialism in favor of what George Handley defines as a "sustained generosity" in light of these passages, an ethical approach that ensures that all of the earth's inhabitants—human and nonhuman alike—can experience the Creation's bounty and beauty (*Hope* 157).

Early Latter-day Saint pioneers understood what it meant to live providently as they established their communities throughout the arid West. They put community first and understood that the water within the streams and rivers was given to them to survive and to help others improve their situations. Widtsoe understood that to flourish in the desert demanded the use of the region's water with judgment so that the community's inheritance was not squandered through selfishness or ignorance. He also understood that successful living also meant that one needed to drink deeply from nature's beauty as a powerful tonic to the soul. Yet his musings on the river are not those of the wilderness lovers seeking isolation and escape on the Colorado reflected in previous chapters. Rather, the ennobling sentiments he feels on the river work in tandem with the mandate to enrich his neighbors, who constitute, in essence, the entire human family. Grand in their scope, stewardship and provident living turn those who embrace these teachings away from oneself and toward the needs of others. In a statement posted on the church's welfare page, Dieter F. Uchtdorf, a current member of the Quorum of the Twelve Apostles, explains, "We all need each other, for it is in sacrificing our time, talents, and resources that our spirits mature and become refined" ("Providing in the Lord's Way"). According to Uchtdorf, such sacrifices make stronger, wiser, and more generous people. In the context of Colorado River use, stewardship and provident living radically reorient how we interact with the river and the basin's diverse inhabitants. They require a recognition that the water is a finite resource, that we must not squander this precious entity, and that we might prepare for hardships ahead so that we and future generations can continue to benefit from the Colorado's life-giving and inspiring waters. Such preparation demands sacrifice and a curtailment of present practices that are not sustainable in the long run. A water ethic of provident living urges us to think beyond our own sphere and time to consider whether we are living beyond our means when it comes to our water bank account or if there will be available water in times of greater need.

This treatment of the water culture of the Church of Jesus Christ of Latter-day Saints demonstrates the vital contribution religion can and should make in shaping society's relationship with water, as expressions of faith speak to the most fundamental questions about the purpose of life and what it means to be part of the human family. As Elder Nash explains, "Faith and religion should have the capacity to stretch, enlarge, and change the human soul beyond self, and to inspire love of God and His creations, to think of others, and to consider the needs of future generations, even to the point of sacrificing personal desires. We need that soul-stretching, for the state of the human soul will directly impact the condition and health of the environment—which, in turn affects our quality of life" (270). If religion can truly bring about this monumental transformation so that people become better stewards of the earth and more committed to care for others now and in the future, then we can begin to see Isaiah's mandate anew.

As I have attempted to show in this chapter, members of the Church of Jesus Christ of Latter-day Saints have in many ways fulfilled the literal interpretation of Isaiah's words. But what of their figurative possibilities? George Handley has suggested, "The notion of a blossoming desert need not imply merely a transformation of the external environment but of our internal cognitive perception" (*Hope* 45). In this regard, I find recent references to the desert-as-blossoming-rose belief by church leaders instructive. In her visit to the Navajo Nation in November 2018, Reyna Aburto, second counselor in the General Relief Society Presidency, reflects on her time with the Diné, a people no strangers to aridity. She explains, "I'm so grateful to be here. My testimony and my faith have been strengthened because I'm here among these special people. And I know that more miracles will happen here and that they will continue to blossom…like a rose in the desert…. The Lord will continue to do miracles through them because of their faith" ("Sister Aburto"). A year later, Elder Gerrit W. Gong, a current member of the Quorum of Twelve Apostles, used this phrase at the G20 Interfaith Forum when addressing the value religious groups bring to their respective communities. Speaking of his own experience, he notes, "In my now-global religious community, our experience includes the spirit of 'making the desert blossom as the rose.' Across the earth in manifold diverse circumstances, members of my religious community promote inner and collective peace; protect and treasure precious natural resources; foster harmonious social cooperation; and invite mutual respect for religious freedom and core moral values" ("Elder Gong"). Similar to Sister Aburto's use, Elder Gong's reference to this long-standing western mantra speaks not of transforming the literal deserts of the West but of reclaiming the goodness of the human spirit.[22]

For those dependent on the Colorado River, Aburto's and Gong's views that the blossoming happening within the desert is one of personal change rather than the material application of water underscore the processes that are now needed for western reclamation in the twenty-first century. Just as Widtsoe noted that Utah's farmers changed when presented with scientific fact, the people within the Colorado Basin can alter their behaviors as they better understand the present state of the river and what future modeling suggests. However, this transformation—what some might call a miracle—will not happen just because the facts suddenly make sense. Indeed, as the challenges the pioneers encountered as they tried to redeem the desert indicate, such transformations are not easy and will occasionally fail. Nonetheless, their history also reveals that resolve, adaptation, and community spirit are essential ingredients to successful arid living. We need such values today more than ever. Yet, as this entire project argues, there is something more that is needed to transform one's behaviors, to bring about—in the words of the faith's Book of Mormon—a "mighty change in your hearts" (Alma 5:14). This change relies on a new awareness of one's place within the watershed and one's commitments to all the basin's inhabitants. It relies on the contributions of poetic discourse captured through the watershed's many stories to filter into hearts and minds feelings of humility, reverence, sympathy, forgiveness of past mistakes, and faith in a more sustainable future. People are a remarkable resource. As hearts and behaviors change, then the desert will indeed rejoice.

Epilogue

Gathering Waters

To use a river without violating its intrinsic qualities will require much of us. It will require our learning to think like a river, our trying to become a river-adaptive people.
—Donald Worster, *Rivers of Empire*

In 1940 western artist and Colorado resident Allen Tupper True painted a series of eight murals adorning the Colorado Capitol rotunda to commemorate the Centennial State's relationship with water. The poem "Here Is a Land Where Life Is Written in Water" by another native son, Thomas Hornsby Ferril, accompanies True's *History of Water* paintings. As one surveys the artwork, Colorado's past and hopes for its future unfold as couplets of Ferril's poetry pay homage to this hydrologic progression from a supposed primitive state to one dominated by the advances of science, technology, and industry. The first mural, *Indian Worshipping Rain*, depicts an American Indian man sitting outside his teepee painting a shield, a thundercloud and lightning bolt emblazoned on it, as thunder and lightning rage around him. Below him, Ferril's words read, "Men shall behold Water in the Sky / and count the Seasons by the living Grasses." From this mural, the others "progress" through the arrival of mountain men, pioneers, and gold seekers to those plowing fields, laying irrigation pipe, and building dams. The final mural offers a glimpse into the future, one dominated by technological advancement as planes, trains, boats, and telephone wires offer the viewers in the foreground a continual scene of motion. A massive Golden Gate–like suspension bridge spans a waterway and vanishes into the setting sun radiating from behind towering peaks

For the Future. Painting by Allen Tupper True. Words by Thomas Hornsby Ferril. Courtesy of Kristal Kraft.

as Ferril's words "Beyond the sundown is tomorrow's wisdom" accompany this optimistic scene (37; True).

These are remarkable paintings, especially when considered in light of the growing concerns about water management in Colorado and throughout the arid West. Although Colorado has been celebrated as the "Mother of Rivers" by retired Colorado Supreme Court justice and poet Greg Hobbs in his tribute to the state's waters, the state is, as True's murals and Ferril's words attest, a place caught in a tenuous balancing act between preserving the state's ecological wonders and meeting the needs of a rapidly growing population. We might ask ourselves, as True's final mural seems to query the viewer, "What is 'tomorrow's wisdom' that will transform Colorado's next chapter of water development?" And if we think more broadly, with Colorado representative of the rest of the Colorado River Basin, how will the watershed respond to the growing uncertainties and complications caused by climate change and an ever-increasing set of demands on the river?

One answer to these questions lies in the murals and poetry themselves. Along with the science, technology, and policy shaping water-resource management, the literary and visual arts have much to offer in terms of the ways in which communities think about and ultimately act in relation to aridity, drought, and water use. Just as True's murals celebrate human ingenuity in reclaiming the West's deserts and making the Colorado's lands "blossom like the rose," Ferril's verse speaks to the challenges of conquering an arid landscape:

> The West is where the water was and is,
> Father and son of old mother and daughter,
> Following rivers up immensities
> Of range and desert, thirsting the sundown ever;
> Crossing a hill to climb a hill still drier. (36)

The juxtaposition of True's triumphant conquest of water with Ferril's recognition of the struggles of the pioneer family to transform the region mirrors the significant body of western American literature similarly preoccupied with the West's arid lands and its frontier legacy discussed throughout this project. The tension created within these two narratives captures the dreams and realities of living in the arid West and creates a productive space by which to imagine the wisdom that will guide the basin's future water ethics.

As I have argued throughout *Tributary Voices*, the prevailing discourses that have shaped the imaginative and material West benefit from narratives that operate across a wide range of genres and discourse communities. They invite contemplation, challenge the status quo, reimagine the future, and outline both theoretical and practical solutions and activities by which to approach the Colorado and its human and nonhuman dependents. Thus, tomorrow's wisdom points toward a greater openness that provides room for these narratives alongside the knowledge derived from the sciences, law, and policy—in essence, a discourse ecology of Colorado River water literacy. Such wisdom also draws lessons from the basin's past. Patricia Limerick argued twenty years ago that the New West could be defined through such ideas as continuity, convergence, conquest, and complexity (*Something* 18–21). Based on developments over the past two decades surrounding the basin states' efforts to mitigate drought through contingency planning and the various research and policy efforts to plan for diminishing precipitation, a fifth *C* seems to have emerged: collaboration. The wisdom in this approach is evident in the rise of numerous watershed groups including the Animas Watershed Partnership, Escalante River Watershed Partnership, San Juan Watershed Group, and Friends of Pinto Creek, which operate to address local needs and the various stakeholders within those smaller basins. Education initiatives such as Water Education Colorado and state chapters of Project WET work to increase citizens' water literacy, breaking down the barriers between the specialized knowledge of water policy and law and the general public. The basin's 2019 Drought Contingency Plans and Minute 323 represent important interstate and binational compromises.

Yet such noble efforts of recent years—and really over the past century as states and communities have worked through the often difficult work of compromise—are in jeopardy. The alliances and goodwill that have prevailed of late are tenuous. Arizona's 2018 increase of its water deliveries from Lake Mead earned the collective ire of the other basin states which felt Arizona manipulated data, thereby compromising established conservation agreements (Loomis). The basin states have also condemned Utah's push for the Lake Powell pipeline to move water from the reservoir 143 miles to the exploding desert metropolis of St. George, threatening litigation in federal court if Utah continues to pursue this project that others throughout the basin fear will drastically affect an already stressed system (Maffly). Without the shared support of the entire basin, such activities raise the specter of contention and chaos that all too many predict as other possibilities for the New West.

The path the basin ultimately chooses will depend on how its citizens view the river and water more generally. As John Fleck observes in the introduction to his recent study of western water issues, "If everyone ignores their own adaptive capacity and simply fights for more, or even fights for the share they've got now in a shrinking system, we are led headlong into conflict, with dangerous results" (6). Such a bleak future manifests in Paolo Bacigalupi's cli-fi thriller *The Water Knife* (2015) that depicts the American Southwest reduced to a desiccated wasteland as climate change has all but dried up the Colorado River. In a not-so-veiled swipe at Patricia Mulroy, the former head of the Southern Nevada Water Authority, Bacigalupi's Catherine Case, the agency's leader, terrorizes Arizona as her mercenary assassins, or water knives, secure by any means possible Colorado River water rights. With other western states walling themselves off from their neighbors and climate refugees from the most drought-stricken areas flooding north toward wetter and cooler climes, the novel posits an incredibly violent future that takes prior appropriation to the extreme.

Another scenario unfolds that rejects this extreme xenophobia and isolationist stance. Here the states forge ahead with a collaborative approach that embraces a watershed stance. Of this alternative Fleck continues, "If instead we recognize our ability to make do with less, and invest in institutions that facilitate water sharing, we can create systems for robust, flexible, and equitable water allocation. Only then can we preserve the West that we all have come to inhabit, know, and love" (7). In this approach, states work together to share in water restrictions, allocate water to habitat preservation, preserve agricultural water, and ensure that marginalized communities receive the water they're long overdue. Of course, such collaboration will not come easily, and there will be difficulties ahead. But the efforts such work will necessitate are far more appealing than the alternative chaos that would ensue as each state fights for its share of the river.

The wisdom needed to usher in a new era of Colorado River management will come from listening to the tributary voices and their respective perspectives, which advocate for a new watershed ethic rooted in community, a sustained commitment to place and the goodlife, reverence, provident living, and restraint. Through their reassessment of the river, its lands, peoples, and all entities forming the larger Colorado River Basin community, these perspectives imagine a new water ethic to guide future management choices. In light of the upcoming centennial anniversary of the Colorado River Compact's signing, we might well imagine a recasting of True's murals informed by the discursive ecology of water literacy and a New Western Rhetoric emanating from the tributary voices considered herein. I see the basin's tribes directly involved in water negotiations and development rather than being relegated to some distant past, with women and other minorities taking on an increasingly larger role in river governance. I see acknowledgments of past and present environmental injustices and a willingness to heal these wounds, collaborative efforts built upon our ability to rhetorically listen to one another to ensure a cleaner and healthier river for all, and a greater spirit of reverence and stewardship for the river and all its relations. And perhaps in the last mural, I observe the artist's celebration of literature and art working alongside technology, science, and policy to create a more sustainable Colorado River. Together, these new murals for a twenty-first-century Colorado River would encourage us to recognize our place within the Anthropocene—this new geologic age dominated by human activity. In these images we are reminded how our decisions have and can continue to negatively impact our water resources and the earth more broadly. But they can also inspire us to resee humanity's relationship within what is and will always be our one and only home.

Toward Tomorrow's Confluences

As a final thought on the connections between discourse, the Colorado River, and the West's future, I take a page from Marc Reisner's introduction to *Cadillac Desert*. On a night flight headed to Los Angeles, the author finds himself staring down at the lights scattered along Utah's Wasatch Front, which punctuate the otherwise black void of the Intermountain West. Pondering the effect of aridity on the region's settlement and the western obsession with water and remaking the desert, he notes, "Westerners call what they have established out here a civilization, but it would be more accurate to call it a beachhead" (3). Despite so many efforts to turn the desert into a garden, the West is largely still a place where different shades of brown dominate the landscape.

This realization was reinforced to me years ago as I boarded a plane in Seattle on my way back to Albuquerque, where I was living as a graduate student. The sun poked through the clouds on this particular winter's morning, bringing warmth and light to a city that I find stunningly beautiful yet too cloudy for my liking. As the plane climbed through the fog to cruising altitude, a view of the Cascade Range and Mount Rainier opened up below me, their glaciated slopes gleaming in the bright morning's sunlight. Thick forest blanketed the mountains and hills, a sea of deep green stretching as far as I could see. In only a matter of minutes, we crossed into the Cascade's rain shadow, and a much more familiar world of barren hilltops and dry washes came into sight. After another hour or so, we crossed over Utah's West Desert and the Wasatch Front and into the Colorado River Basin. Snow hardly covered the Wasatch, which are typically buried deep below winter's snows that time of year, an ominous sign for a long, dry year ahead. Soon the canyon country came into sight and then, to my great satisfaction, the Green River.

From thirty-eight thousand feet the canyons opened onto the earth like millions of capillaries, branching every which way as they dug deeper into the land through gravity's relentless pull. The jumbled cliffs of the San Rafael Swell were made minuscule, the great plateaus and buttes south of Green River, Utah, looking like a child's sandbox creations. Eventually, the Colorado came into sight, an equally serpentine river carving its way around and through billions of years of geologic history. Moab, Utah, and the La Sal Mountains were visible to the east. And then, as we continued on our southeasterly course, I saw it: the confluence. Awkwardly twisting and craning my neck to get a better glimpse through the window, I could barely see where the Green and the Colorado meet. Miles away from the nearest town and where an eons-old ritual continually plays out, their waters commingled and moved downriver on their ancient course to the sea.

Having lived the bulk of my life in areas that draw on the river and traveled extensively throughout the watershed, I have spent many years reflecting on the significance of the Colorado. Visiting the headwaters in Colorado and Wyoming, many of the diversions moving water through the Continental Divide to the Front Range, the dams of the Central Utah Project, which transfer Colorado River water into the Great Basin, Lakes Mead and Powell, Las Vegas, the extensive irrigation projects of the Imperial and Mexicali Valleys, and the delta's dry riverbed and few wetlands, there is no doubt as to the Colorado's influence in this region. Yet when hurtling through the sky tens of thousands of feet above the ground at more than five hundred miles an hour, the Colorado seems quite insignificant. The river and its tributaries appear as tiny, sinuous threads weaving their way through millions of acres of mountains, valleys, and deserts. With so much of the watershed void of significant greenery despite all the efforts to reclaim the desert, reality seems to surpass any belief that this place could be entirely redeemed. From my vantage point soaring high above the desert interior of the Colorado watershed, I realized, as Reisner had decades earlier, that John Wesley Powell was prophetic in his observations about the West.

Now, after years of immersing myself in the history and literature of the Colorado River country, I have also come to see the wisdom and insight of more recent explorers and inhabitants of the basin such as Craig Childs, Patricia McCairen and Evelyn White, the Nuestro Río coalition, members of AEURHYC, Natalie Diaz, Leslie Marmon Silko, Ofelia Zepeda,

Nikki Cooley, John Widtsoe, and so many others. While a physical redistricting of the basin's geopolitical boundaries is nigh impossible, the watershed visions these voices promote can nonetheless redraw the map imaginatively along more just and ethical lines to promote a healthier river system and more adaptive, humble river users. With the fate of so many in the United States, Mexico, and tribal nations resting on an overallocated, highly regulated, and stressed river, perhaps our greatest resource in addressing the Colorado's challenges is the shared commitment to see the river flow onward toward its age-old confluences.

Notes

INTRODUCTION

1. The Colorado River Basin includes Arizona, California, Colorado, Nevada, New Mexico, Utah, and Wyoming as well as the Mexican states of Baja California and Sonora.

2. Since the construction of the large-scale dams on the mainstem of the Colorado, the flows to the delta have greatly diminished. I discuss these impacts and the more recent restoration efforts in chapter 4.

3. The 2007 Colorado River Interim Guidelines specify that if the reservoir elevation at Lake Mead drops to or below 1,075 feet, the secretary of the interior declares a shortage to begin reducing deliveries to the Lower Basin states.

4. "Hot drought" refers to drought conditions exacerbated by higher average temperatures, which affect snowmelt, soil moisture, and evapotranspiration rates.

5. The full report is titled "Record of Decision: Colorado River Interim Guidelines for Lower Basin Shortages and the Coordinated Operations for Lake Powell and Lake Mead." Reclamation released a review of the guidelines in December 2020; they are set to expire at the end of 2025.

6. The program has since been rebranded under the bureau's current WaterSMART Program. See Ogren and Wolf 13–16.

7. Some reclamation leaders believed that the Colorado's average annual flow could support 20–22 million acre-feet of development per year, while the 1922 compact eventually settled on 17.5 maf/year. However, through their analysis of E. C. LaRue's studies of the river's flows, Kuhn and Fleck show that there was far less water in the river system than many wished to believe, suggesting that a more realistic projection was 13.5 maf based on records from 1886 to 1905 (Kuhn and Fleck 33, 64, 76, 212–13).

8. The Green River's previous names included the Buenaventura and Seedskeedee, among others, representative of the Spanish and Native presence in the area.

9. See Wikipedia's "Colorado River" entry, Larry Anderson's "Utah's Perspective: The Colorado River" report for the Utah Division of Water Resources, and Sarah Zielinski's *Smithsonian Magazine* article "The Colorado River Runs Dry" as just a few examples of this point. Jonathan Waterman's *Running Dry* (17) and Luke Runyon's "How the 'Grand' Became the 'Colorado'" also provide a good discussion of this naming issue.

10. To Congress Taylor remarked, "It has always seemed a strange and anomalous situation that our great State should be named for one of the greatest rivers in the world that arises and flows hundreds of miles within our borders, and that the river should bear the name of our State from its mouth clear up nearly to the border of the State and then be called by some other name within our state.... We have in Colorado no State pride in the name 'Grand.' That name is merely an adjective, and does not mean anything but

large or great, and might be applied to any large or beautiful stream. Practically everything in Colorado is grand, and as applied to a river it is a very commonplace name. It has no historical or special signification, while on the other hand the magic word 'Colorado' does have a very great and affectionate significance to every civilized human being on this planet.... [F]or the past 60 years, 'Colorado' has meant the heart of the Golden West, the actual top of the world, the land of sunshine, good health, and gorgeous scenery, the summer playground of the Nation, the Switzerland of America, the bright jewel set in the crest of this continent, where it shines as the Kohinoor of all the gems of this Union; the sublime Centennial State" (8).

11. Colorado received roughly 3.8 maf, whereas Utah received 2.8 maf, Wyoming 1.0 maf, and New Mexico 0.8 maf.

12. Wilkinson identifies the other "lords of yesterday" as the 1872 Hardrock Mining Law, policies regulating the nation's public domain, and those impacting the Pacific Northwest's salmon runs.

13. Nash looks to the European Romantics like Burke and Kant and their musings on the sublime as the foundation for an appreciation of wilderness. For a more specific discussion of this topic, see Nash's chapter "The Romantic Wilderness."

14. This framework takes its cue from Spinuzzi and Zachry's genre ecologies, a model developed to understand workplace communication and how "human interactions with complex technologies are inevitably mediated by dynamic and unpredictable clusters of communication artifacts and activities" (170–71). I am indebted to Michelle Kells for this consideration of genre ecology and its relation to water literacy.

15. Perhaps not so ironically, their "myths" have been discounted, whereas the West's settler-colonial culture has embraced wholeheartedly such stories as "rain following the plow."

CHAPTER ONE. "HUMANISTS AT THE HEADGATES"

1. The question of the efficacy of scientific fact in motivating behavioral changes is a key issue for many scholars. As DeLoughrey, Didur, and Carrigan point out, the "simple insistence on the scientific facts remains politically ineffective when it is disconnected from the political, social, cultural, affective, and rhetorical forms that the climate problem takes in different communities" (3). Steven Allison and Tyrus Miller further corroborate this point: "So far, scientific facts have not motivated Americans to support the huge societal transformations needed to stop climate change. Some reject the scientific consensus on global warming because it makes them feel bad or clashes with their personal experience of the weather." They see action resulting from issues wherein climate change has more direct and visible impacts on "people's homes, livelihoods and spiritual beliefs" ("Why Science Needs the Humanities").

2. Heise explains that the environmental humanities, as a specific field, emerged in the first decade of the new millennium and has since spread across the globe, building on the contributions of earlier established disciplines such as environmental philosophy, environmental history, and ecocriticism (1).

3. As environmental crises become ever more complicated and the knowledge addressing them more specialized, the environmental humanities as a broad discipline now

houses more focused subgroups. Some of these include the Arctic humanities ventures at the University of Lapland and Boston University and the Desert Humanities Initiative at Arizona State University.

4. Limerick suggests that Turner's belief that the frontier had closed has merit only if viewed through the effects of agriculture on the region. On the other hand, she notes how the notion of the frontier continued to live well beyond 1890 in industries like mining and ranching. See *The Legacy of Conquest* 23–24.

5. Worster's call to employ Powell's model has gained significant traction in recent years, as numerous environmental thinkers and water managers have looked to the watershed as a useful structure to delineate water use. Jack Loeffler and Celestia Loeffler's book *Thinking Like a Watershed* builds on Aldo Leopold's mantra of "thinking like a mountain" and Powell's vision to posit new directions for living with rivers. Similarly, an April 2011 Carpe Diem West report on Colorado River Management titled the work "Thinking Like a River Basin."

6. The notion of the local has long been seen by environmental advocates as a panacea to current issues. Such has been the case within ecocriticism. However, more recent scholarship has questioned its value, evident in such seminal works as Ursula Heise's *Sense of Place and Sense of Planet*. For a further discussion of this debate between the local and global, see Lynch, Glotfelty, and Armbruster's *The Bioregional Imagination*; Lynch et al.'s *Thinking Continental*; and Richard and Formisano's "It Was the River" in *Reading Aridity in Western American Literature*.

7. Fullan's hydromythologies shed further insight on my framework wedding the watershed to the discourse map. For her, a hydromythology "is a story-map that leads to meaningful actions, and shapes the relationships between people and water in the cultural context from which it emerges and to which it contributes, and the story-map overlays the physical and spiritual relationships that arise to and out of specific places" (137–38). Thus, the tributary voices' narratives or hydromythologies operate across various temporal, spatial, discursive, and emotional boundaries to generate new meanings and behavior.

8. Ratcliffe clarifies that traditional notions of identification seek to establish common ground but, in doing so, create a problematic situation, as "differences are often glossed over or erased" (32).

CHAPTER TWO. SCIENCE, POETRY, AND PARADOX

1. Their studies examine contributions by individuals such as Cabeza de Vaca, John C. Frémont, William Manly, Mark Twain, Mary Austin, John Van Dyke, Joseph Wood Krutch, and Edward Abbey, among others.

2. The little scholarship that exists on Childs's writings places the entanglements of scientific and aesthetic epistemologies at the forefront. Matthew Teorey has suggested in his study of this text that Childs's work "is less about sharing scientific data and recounting exciting backpacking adventures than it is expressing an ecocentric outlook, whereby readers begin to comprehend and grapple with humanity's eco-reality and biological selfhood" (3). To facilitate this new vision of nature and society's relationship to it, Teorey argues that Childs employs an "ecological discourse, an articulation of

nature's intrinsic value and humanity's responsibilities as a member of the biotic community" (2). Although I agree with Teorey's assessment of *The Secret Knowledge of Water* as a means of reorienting the reader's mind to a more ecocentric perspective, I find that Childs's scientific training, awareness, and scientific discourse, which speak to his motivation to know the desert and its water, play an integral role in defining his unique vision of these entities.

3. Kino's travels covered primarily southern Arizona and northern Mexico in search of converts. While proselytizing was Kino's primary objective, he was nonetheless aware of the opportunity for the empire to gain wealth in the area. In his letter to Spain's viceroy in Mexico City in 1703, he writes, "It would be no difficult task to acquire very rich valleys watered by bounteous and fertile rivers, and densely populated by well-disposed and docile natives, long inured to work. Nor to be overlooked are the mining possibilities. All this wealth could be a source of profit and renown to the royal Catholic dynasty and to our mighty monarch, Philip V" (24). Franciscans Domínguez and Escalante left Santa Fe in 1776 searching for a route to the mission in Monterey, California. Although they never reached the mission, their wanderings through present-day New Mexico, Colorado, Utah, and Arizona provided valuable accounts of the Colorado River and surrounding country still celebrated today.

4. "Topos" and its plural, "topoi," derive from Aristotle's work on understanding rhetoric, or how one forms arguments to persuade an audience. "Topoi" refer to commonly held assumptions, ideas, and beliefs to build an argument. The following chapter traces more closely how women writers use these conventions for varying rhetorical purposes.

5. Pike believed the Mississippi and Missouri Rivers would confine western expansion, leaving "the prairies, incapable of cultivation, to the wandering and uncivilized Aborigines of the country" (231). As Goetzmann argues, Pike's description of the Louisiana Purchase's southern territory facilitated the myth of the "Great American Desert," predating Stephen Long's views on the region to which most scholars have attributed the idea (51).

6. The scientists on the steamship included Balduin Möllhausen, a relative of Humboldt.

7. Two years prior to Austin's publication of *The Land of Little Rain*, eastern art critic John C. Van Dyke published *The Desert*, an account of his travels through the Mojave and Colorado Deserts of southern Nevada and California. While his text is one of the most famous desert texts, it has also received much scrutiny in recent years for his inaccuracy and fabrication. Peter Wild's introduction to the 1999 American Land Classics edition of this text provides a thorough overview of Van Dyke's impetus for writing and reveals the truth stretching of the author's "authentic" observations (Introduction). I have chosen Austin's account as a more indicative model of the characteristics of nature writing that Slovic addresses.

8. Abbey's chapter "Cliffrose and Bayonets" from *Desert Solitaire* aptly describes his own reliance on this motif. Leaving his trailer one early winter morning, he announces, "Time to inspect the garden" (27). Describing the expanse of land that denotes his little piece of heaven, he goes on to "inventory" the rich variety of flowers, shrubs, and cacti that add to the beauty of his surroundings (27).

9. Childs published a more detailed version of this chapter in the spring 2009 issue of *Wings*, the official publication of the Xerxes Society for Invertebrate Conservation. It is titled "The Memory of Water: Life in Ephemeral Water Holes" and located on Childs's author website, House of Rain.

10. Extensive livestock grazing and other activities in the desert Southwest have damaged huge areas of cryptobiotic soils, exacerbating spring dust storms that ultimately have negative effects on the region's water resources and on human health (Deems 4401; MacMillan).

11. In the chapter "Chubasco," Childs also notes how his mathematical calculations fail to account for the unpredictability and unknowability of water. Rationalizing his presence in a deep canyon amid a downpour during a month when Arizona flash floods killed nineteen people, Childs explains, "I had made choices at the knoll, calculations on where water would flow, how much, how quickly. I had forgotten the seventeen lines of hard mathematics that had proven not a thing about desert rain" (*Secret Knowledge* 204).

12. According to Donald Worster, Whewell was the first to use the term "scientist" (*Nature's Economy* 130).

CHAPTER THREE. NEW CURRENTS IN COLORADO RIVER BOATING NARRATIVES

1. More recently, Nina Baym sarcastically opens her extensive study *Women Writers of the American West, 1833–1927* asking, "How could there be books by women about the American West, when everybody knew that the topic was reserved for male authors?" (1).

2. Norris Yates argues that Owen Wister looked to Mary Hallock Foote's *The Led-Horse Claim: A Romance of a Mining Camp* (1883) as one of the primary inspirations for *The Virginian* (11).

3. At the 2015 annual conference of the Western Literature Association, Comer admonished the audience to take more stock of women's contributions to our understanding of the West, arguing that despite such gains over past decades, there is still considerable work to be done to recover and foreground their work.

4. Weiler Walka writes about experiences on the San Juan River. Meloy's and Hansman's works describe journeys along the Green River, whereas the others focus on the Colorado's mainstem. Christa Sadler's *There's This River: Grand Canyon Boatman Stories* (2006) is an important collection that provides a unique glimpse into the world of the canyon's river guides. About 20 percent of the book includes stories by women, but the collection is less concerned with gender than with the stories emanating from the canyon. Yeti, an outdoor equipment company, released a short video, *In Current*, about Amber Shannon and her dream to guide dories, small wooden boats, through the Grand Canyon.

5. Fedarko's text is less attuned to his own experiences on the river than the other texts listed. Nonetheless, his masterful account of the history of Grand Canyon dories and Kenton Grua's fastest-known time rowing the Grand Canyon in 1983 securely places this highly entertaining and informative book within this genre.

6. As Jane Tompkins contends, language and the western hero are at great odds with one another, as the expression of language represents everything the hero hopes to resist. She explains: "Westerns distrust language. Time and again they set up situations whose message is that words are weak and misleading, only actions count; words are immaterial,

only objects are real.... Doing, not talking, is what it values" (49–50). Thus, in Abbey's opening scene where he and his friend Newcomb load their boats, Newcomb is silent, alone with his own unarticulated thoughts, while his actions, the smoking of his pipe and the work he performs, do the speaking in this scene. After a nonsensical exchange between the two following Abbey's inquiry to understand where they come from, he concludes, "Words fail" (*Desert Solitaire* 202). Indeed, words do fail here as the men float onward down the river, yet the meaning of this and the opening scene is clear. Language isn't needed to describe what is happening since its absence suggests all we need to know about the point behind this voyage. Tompkins reminds us of the Western's purpose and notes its influence on how we perceive gender: "Because the genre is in revolt against a Victorian culture where the ability to manipulate language confers power, the Western equates power with 'not-language.' And not-language it equates with being male" (55).

7. Dellenbaugh penned the introduction to North's account, *Down the Colorado by a Lone Girl Scout, Mary Remsen North*.

8. Nevills was married to Norm Nevills, a renowned Colorado River boatmen and leader of Clover and Jotter's trip. Doris Nevills and Mildred Baker were the first women to boat from Wyoming to Arizona (Webb 133).

9. Clark swam sections of the river in 1945 and 1946 with friend Harry Aleson. In *Georgie Clark: Thirty Years of River Running*, she explains, "I wanted to make the river accessible to everyone regardless of age, sex or physical condition" (90). Her pioneering efforts in building motorized rafts made this desire a reality.

10. This vignette provides a corollary to the events outlined in Fedarko's *The Emerald Mile* as they relate to the historic 1983 runoff that caused significant damage at Glen Canyon Dam, created some of the largest rapids ever recorded along the Colorado, and ensured the fastest-known time paddling through the Grand Canyon.

11. In *Legacy of Conquest*, Limerick problematizes the historical roles in which western women have often been cast, arguing that the trope of women as "innocent victim" or the fallen woman is far more complicated when looking at individual women's lives. Limerick concludes, "Prostitutes were not consistently and exclusively sinners, nor were wives and mothers consistently and exclusively saints" (54).

12. "For the really strong man, language is a snare; it blunts his purpose and diminishes his strength" (Tompkins 51).

13. Hogan's short essay highlights the datura flower and uses it as a means of reflecting on the river's Indigenous presence. She speaks of "another history than the ones recounted here by our river guides, a history that begins only with the journeys of white men" (64). Hogan goes on to explain, "This new history" fails to account for river histories of the Havasupai and Hopi and the ways in which they construct "relationships with all other life forms" such as the sacred datura (64, 67). As these multispecies relations are part of the focus of chapter 5, I develop the ideas expressed in Hogan's writing in greater detail there. Suffice it to say, her contribution to *Writing Down the River* challenges not only the exclusionary record provided by the "journeys of white men" but the overall racialized vision of the river and canyon that has historically obscured the profound Indigenous connections to these places.

CHAPTER FOUR. *NUESTRO RÍO*, THE GOODLIFE, AND *QUERENCIA*

1. Even Daniel Beard, a former Bureau of Reclamation commissioner, writes, "To visit Glen Canyon is to undergo an awe-inspiring experience. It is truly a magical place and to continue to flood this natural wonder is a crime of monumental proportions" (74).

2. I use these terms to differentiate among the many different Spanish-speaking communities in the Americas. I follow Frances Léon Quintana (Swadesh) who defines Hispano as one who is Spanish speaking and a "descendant of seventeenth and eighteenth century colonial settlers of the [American Southwest]" (*Los Primeros Pobladores* xviii–xix). "Latina/o" represents a more general description of Spanish speakers in the American Southwest but with ancestral ties to Mexico, while "Mexican" refers to citizens and residents of Mexico. Throughout this project, I distinguish between these groups when addressing specific texts and histories. Otherwise, I employ "Latina/o" when referring to the American Southwest's Spanish-speaking cultures generally.

3. See Richard and Formisano's "'It Was the River': Indigenous Anti-dam Literature of the Great American Desert" for discussion of Indigenous anti-dam texts.

4. El Vado de Los Padres—the Crossing of the Fathers—in Glen Canyon was where the Spanish friars passed through on their return trip to Santa Fe. Near the Colorado's confluence with the Escalante River, Mormon pioneers crossed through Glen in 1880 on their way to colonize southeastern Utah as part of the San Juan Mission. Martin notes that the Crossing of the Fathers was a regular point of passage among the region's tribes (115). Glen Canyon Dam was built on Navajo Nation lands, and sacred sites like Rainbow Bridge were impacted by the dam's construction, the rising of the reservoir, and the influx of tourists to the region (Farmer 91). While government agencies have worked to address the historic and current tribal connections to the canyon evident in the Glen Canyon Dam Adaptive Management Program, mainstream considerations of Glen Canyon's disappearance tend to focus on wilderness loss rather than cultural compromises ("Tribal Resources").

5. The Hispano settlement occurred as a result of the Brunot Agreement in 1874 that opened up Ute lands for non-Indian settlement and eventually removed the various Ute bands to reservations in southwestern Colorado and eastern Utah. Frances Quintana explains that the Hispano settlers in the region opposed their removal, as they feared "these measures would strengthen the position of the Anglo population at the expense of both the Hispanos and Utes" (*Los Primeros Pobladores* 98). Here we see how "Latinxs inherit genealogies of both colonizer and colonized" (Wald et al. 7).

6. Quintana gave up the name Swadesh after her divorce and remarriage to Miguel Quintana in 1978. Her papers at the University of New Mexico's Center for Southwest Research are listed under Quintana. My future references to her book *Los Primeros Pobladores* will use Quintana to avoid switching between the two names. I also list her book under Quintana in the works cited, although it was originally published under Swadesh. Another resource on the dam's impact is Leonard Trujillo's 2006 interview as a Bureau of Reclamation employee. Trujillo was born below the current dam site and eventually became reservoir superintendent. He recalls growing up in Los Martinez, "a real nice community where Navajo Dam is at, below it" (2). Most of the interview is

about his career with the bureau, but he does recall fond memories of attending school in Los Martinez and working the acequias there.

7. I find it somewhat ironic—but not surprising based on Nixon's own assessment of American anti-dam literature—that his chapter on Rosa is framed within the context of traveling to the San Juan's tributaries to fly-fish gold-medal trout streams. While this approach is appropriate within the larger purpose of his work, it nonetheless relies on various tropes associated with wilderness narratives such as the lone man seeking refuge from the city and nostalgia for some lost Eden. However, Wilkinson offers readers an important glimpse into the issues surrounding Rosa's demise, providing more firsthand accounts than Quintana's study. My point here isn't to fault Wilkinson, but to address the complicated and often competing ways in which people represent the river.

8. Nichols's novel is based on the bureau's failed attempt to build the Indian Camp Dam in the 1970s as Taos, New Mexico, locals challenged the organization and the dam's proposed impacts on the local water culture. See Marilyn Koch's history of the Indian Camp Dam debates and their connections to Nichols's work.

9. Barry Lopez traces this term to its roots in the tradition of Spanish bullfighting, something that Ernest Hemingway extensively elaborates on in *Death in the Afternoon* (1932) wherein he explains that *querencia* is "a place the bull naturally wants to go to in the ring; a preferred locality." It is here that "[the bull] feels that he has his back against the wall and in this *querencia* he is inestimably more dangerous and almost more impossible to kill" (150). Arellano's scholarship on *querencia* has inspired the recent collection by Fonseca-Chávez, Romero, and Herrera titled *Querencia: Reflections on the New Mexico Homeland* (2020).

10. As of December 2020, the *Nuestro Río* website no longer exists. Other media promoting the Latina/o–Colorado connection include American Rivers' Spanish version of their 2013 *I Am Red: The Colorado River* short film, *Yo soy rojo: El Rio Colorado*, produced in 2015, and Clemente Álvarez and Nacho Corbella's *Colorado River: Hispanics to the Rescue* project for the Univision Planeta series.

11. *Nuestro Río* includes a link to this short film on its Facebook page.

12. The film portrays a very optimistic representation of labor conditions in the Yuma area. Other than Gonzales's brief mention of the difficult work he performs, the film does not address issues of safety, migrant labor, or poor wages, among other labor-related concerns. Nonetheless, the film offers an important overview of some of the values that shape those living and working in this part of the Colorado River Basin. It's worth noting that César Chávez, the iconic labor activist, was born in Yuma and has been an influence for Nuestro Río and its mission to connect Latina/o peoples to their Colorado River heritage. See Troy Hooper's reporting in the *Colorado Independent* for more on this relationship.

13. Leopold ends "Green Lagoons" with this plaintive query: "All this was far away and long ago. I am told the green lagoons now raise cantaloupes. . . . Man always kills the thing he loves, and so we the pioneers have killed our wilderness. Some say we had to. . . . Of what avail forty freedoms without a blank spot on the map?" (157–58).

14. The Colorado Delta has inspired numerous writers to attempt to capture this unique place. For more on the delta's literary representations of the region, see my chapter in Paula Farca's collection *Make Waves: Water in Contemporary Literature and Film* (2019). Over the past two decades, various media have focused on the environmental degradation that has greatly altered the area as a result of the expansive agricultural, urban, and infrastructural developments throughout the United States and Mexico. Some of the more notable works include Bergman's *Red Delta* (2002), Jonathan Waterman's *Running Dry* (2010), and David Owen's *Where the Water Goes* (2017), as well as short films including *Death of a River: The Colorado River Delta* (2010) and *Renewal* (2014). Of course, there is a substantial body of scientific reports produced by US and Mexican scientists and organizations that have contributed knowledge about the delta's threatened ecosystems. For a more detailed list of sources, see the University of Arizona's Colorado River Delta/Delta Del Rio Colorado Research Coordination Network website. It includes links to online bibliographies, websites, books, and other academic publications about the delta.

15. US-Mexico treaties regarding the Colorado River are tied to the two nations' treaties on the Rio Grande. While Mexico has the lion's share of that river's water, the United States takes more of the Colorado. See Norris Hundley's *Dividing the Waters* for more on the Rio Grande's impact on Colorado River deliberations.

16. Reflecting the Mexican perspective during these years, a relative of President Manuel Ávila Camacho lamented, "The cotton will be lost if our 'good neighbor' don't loosen water from the Colorado River. These gentlemen are our 'good neighbors' since 1847 and they either make war on us or drag us into it according to their desires" (qtd. in Ward 25). Such distrust and displeasure among Mexicans caused President Lázaro Cárdenas to seize all foreign-owned lands in the Mexicali Valley in 1937 (5). Heightened tensions continued to characterize the US-Mexico relationship over water throughout the first half of the twentieth century as each nation refused to acquiesce to the other's demands regarding the Colorado, in Mexico's case, and the Rio Grande, in the United States' (Hundley, *Dividing the Waters* 73–74).

17. This was previously known as the International Boundary Commission, formed in 1889 to resolve border disputes and address the ever-shifting border formed by the unpredictability of the Colorado's and Rio Grande's flows. See the 1944 treaty, Article 2, for more information.

18. The question regarding whether a Mexican equivalent to Abbey exists is a provoking one, as the literature from the region does not reveal such an author as far as I have been able to discern. What I find most significant in this question is not so much whether one exists but why one should ask the question in the first place. Such a query reveals a conditioned view on the types of voices that can and should address environmental issues. In the case of Leopold and other prominent American writers of the Colorado River watershed (as evidenced in chapters 2 and 3), a prowilderness rhetoric typically emerges as authors praise a place for its wildness, isolation, and human absence. Such perspectives clearly reveal a bias for particular locales and the privileged background of those able to retreat to such places. Thus, by asking this question we assume

that the delta is a wilderness, a place that inspires writers to reflect on the loss of such places. While such work can be useful in drawing awareness to a locale, it usually fails to acknowledge those who consider the place home rather than a leisurely retreat in which to lose oneself.

19. "Trabajando para la Conservación del Delta del Río Colorado." All translations are mine.

20. "Conservar y restaurar los ecosistemas del Delta del Río Colorado y promover un buen uso de sus recursos para la realización y permanencia de actividades que permitan el desarrollo economico, social, y cultural de las comunidades locales" (7).

21. "Quedamos pocos de los que vimos al Río Colorado en su explendor, antes de que su flujo cesara casi por completo. La mayoría de nosotros, ahora miembros de AEURHYC, tuvimos nuestro primer encuentro con el Río Colorado en la década de los 1950s, y algunos otros, como miembros de la tribu Cucapá, desde hace cientos de años" (2).

22. "Desde entonces hemos sido testigos de cómo el río ido deteriorando, y como esto ha afectado nuestras vidas" (2).

23. These texts include Lentfer and Servid's *Arctic Refuge: A Circle of Testimony* (2001), Moore and Slovic's *Wild Nevada: Testimonies on Behalf of the Desert* (2005), and, most recently, Trimble's collection *Red Rock Stories* (2017)—based off of *Red Rock Testimony* (2016), which was delivered to Congress to promote the creation of Bears Ears National Monument. As with *Testimony, Arctic Refuge* and *Red Rock Testimony* were presented to members of Congress to urge greater wilderness protection in Alaska and Utah.

24. "Disfrutar de el como alguna vez nosotros lo hicimos" (2).

25. "Recuerdo que en 1952 se abatió un chubasco de verano que inundó todo" (2).

26. "Las aguas del río fueron nuestro único alivio durante los largos días de verano" (2).

27. The floods had a particularly devastating effect on Campo Mosqueda, a popular ecotourism site along the Rio Hardy. See *Red Delta* 194–97 for a detailed account of the flooding and the resulting lawsuit filed by Jesús Mosqueda González against the United States for damages.

28. "Da mucha tristeza ver que lo que era un caudaloso río se haya convertido en un sucio dren que en partes solo alcanza cinco metros de ancho" (3).

29. A good example of embedded rhetoric is seen in Childs's writing of desert water, where politicized viewpoints about how to live in the desert are deeply embedded within the "epistemological inquiry" that defines his writing (Slovic, "Epistemology" 90).

30. The groundwork for Minute 319 came from Minute 306, signed in 2000, which identified preliminary efforts by a small group of interested parties on both sides of the border to address the delta's ecological health. It also mandated binational collaboration to see the ongoing engagement by a greater number of participants to work toward restoring the region ("Minute 306" 1). See Raise the River Coalition's short film *Renewal: A Reborn Colorado River Once Again Finds Her Path to the Sea* (2014), which captures the elation of seeing a free-flowing river south of the border once again.

31. The report notes that in 2014, there was a 17 percent increase in vegetation growth and overall rise in the water table. However, the positive results diminished over time the closer one came to the sea. See "Minute 319 Colorado River Limotrophe and Delta Environmental Flows Monitoring Final Report" 17–19 for a summary of results.

CHAPTER FIVE. INDIGENIZING WATER JUSTICE ALONG THE COLORADO

1. H. M. Valentine, assistant secretary-treasurer of the Navajo Trail Association, explains in a press release that Navajo Route 101 would save 260 miles between Los Angeles and Chicago and make travel between such national parks as Grand Canyon and Mesa Verde a half-day affair.

2. The landmark *Arizona v. California* ruling offers further clarification on tribal water rights, as it quantified allotments to Lower Basin tribal reservations based on early priority dates including the Chemehuevi, Cocopah, Yuma, Colorado River, and Fort Mohave Reservations.

3. A day prior to Jones's letter to Watkins, Jones testified before Congress to reaffirm the Navajo Nation's commitment to the Navajo Unit. Addressing New Mexico representative Antonio Fernandez and the House Speaker, Jones emphatically states his desire to see the San Juan River "harnessed, controlled, and used" and reminds his audience to whom the water belongs and where the federal government stands in regards to these reserved rights: "A part of it belongs to the Indians and no Congress is going to take it away from them" ("Congressional Record").

4. As the Bureau of Reclamation's history of the San Juan–Chama Project notes, interest in irrigating northern New Mexico from the San Juan surfaced as early as the turn of the twentieth century. The idea of a transbasin diversion between the San Juan and Chama Rivers emerged in the 1930s ("San Juan–Chama Project" 6–7). Today, the project captures water from a number of San Juan River tributaries in Colorado and diverts it through the Azotea Tunnel under the Continental Divide. From there, the water moves into Heron Reservoir, where it eventually flows into the Rio Chama and finally the Rio Grande.

5. Tribes are recipients of some of the San Juan–Chama water, including the Jicarilla Apache and various Pueblo along the Rio Grande.

6. Much of the funding and construction holdup had to do with the ongoing negotiations between the Bureau of Reclamation, the Bureau of Indian Affairs, the Navajo Nation, and other entities, including the Jicarilla Apache Nation, about how to carry out this extensive project (Glaser, "Navajo Indian" 17–23).

7. For more on the notion of the ecological Indian, see Shepard Krech's *The Ecological Indian: Myth and History* (2000).

8. Adamson and Monani ground their discussion of Indigenous cosmovisions within the theory of cosmopolitics, the origins of which they trace in the introduction to their edited collection *Ecocriticism and Indigenous Studies*. They conclude that cosmopolitics "implies that we are entering a moment in politics that takes as its goal... the recognition of intergenerational, evolutionary space and time required not just for the survival of all species, but for the recognition of the 'rights' to life for all humans and nonhumans" (7). Looking to Adamson's extensive work in this emerging perspective, Mascha Gemein defines ecopolitics as a "theoretical framework and political strategy for an expansion of political participation beyond the human realm" and notes that it "corresponds to the philosophies guiding Indigenous movements" (485).

9. Such are the issues at the heart of the recent Dakota Access Pipeline protests on the Standing Rock Reservation. In *Our History Is the Future* (2019), Nick Estes cites the

Lakota and Dakota idea of Mitakuye Oyasin, or "all my relations" or "we are all related," in his treatment of Indigenous efforts to protect the Mni Sose (Missouri River) (15).

10. Kang identifies such genres as "the historical epic, literary realism, Victorian children's literature and women's botanical writing" as those through which Silko illuminates the social and environmental racism that has shaped and continues to confront Indigenous experience in the region (103).

11. While Silko situates the novel's action at the turn of the century, construction on the Colorado River Aqueduct Parker Dam did not begin until 1933, with Parker Dam construction beginning a year later and completed in 1938.

12. According to a Department of the Interior press release in March 2006, Secretary Gale Norton signed the Southern Arizona Indian Water Rights Settlement allocating 37,000 acre-feet to the Tohono O'odham Nation to be delivered from the Central Arizona Project, the state's massive water-distribution project that pumps Colorado River water hundreds of miles to the Phoenix and Tucson areas ("Secretary Norton").

13. "How the Milky Way Was Made" first appeared in her poetry collection *When My Brother Was an Aztec* (2012).

14. In his opening remarks at a 2018 water law symposium, Ernest House Jr., a member of the Ute Mountain Ute Tribe and former executive director of the Colorado Commission of Indian Affairs, shared wisdom his father had passed on to him after years of representing the tribe: "If you're not at the table, you're on the menu" ("Enhancing").

15. The Ten Tribes Partnership represents only a portion of the tribal interests in the watershed, as there are twenty-nine reservations served by the Colorado River (*Colorado River Basin Ten Tribes Partnership*). See the "Colorado River Basin Tribes" page on the tribalwateruse.org website for more information about the tribes in the watershed.

16. Adamson and Monani look to Grace L. Dillion's *Walking the Clouds: An Anthology of Indigenous Science Fiction* (2012) to formulate their assessment of Indigenous scientific literacies within the broader theories of cosmopolitics and cosmovisions. Lejano, Ingram, and Ingram remind us that the Western divide between science and other forms of knowledge does not define Indigenous epistemologies: "Indigenous or traditional knowledge can combine observations of physical phenomena and symbolic and cultural meanings. Traditional knowledge networks have fiercely resisted the division of knowledge along the lines of what is more or less acceptable to dominant culture" (13).

17. The Colorado Plateau Foundation works with Indigenous communities in protecting local ecosystems, working toward food security, and preserving Indigenous languages.

18. The Escalade Project envisioned a large resort at the confluence of the Colorado and Little Colorado Rivers, complete with lodging, dining, and shopping options, as well as a tram that would carry up to ten thousand people per day from the rim to the confluence ("Stopping Grand Canyon Escalade"). Pumped Hydro Storage has plans for four dams along the Little Colorado, whose reservoirs would inundate numerous sacred sites throughout the canyon (C. Nelson, "This Land Is All We Have Left").

19. In April 2019, Senator Mitt Romney introduced Senate Bill 1027, the Navajo Utah Water Rights Settlement Act of 2019, to quantify the nation's water-rights claims and provide funding to bring water infrastructure to many of the reservation's homes in Utah that lack running water. Congress passed the bill in December 2020.

20. Forty percent of homes in the Navajo Nation have to haul water. The Navajo-Gallup Water Supply Project seeks to improve water infrastructure and access for those on the eastern side of the reservation, portions of the Jicarilla Apache Reservation, and the city of Gallup. As of 2020, the federal project has begun to deliver water, and all project construction is required to be completed by 2024 ("Navajo-Gallup").

CHAPTER SIX. RECLAIMING RECLAMATION

1. This is the official name of the church, unlike Mormon, the more well-known nickname often used to describe the faith and its members. The term "Mormon" is not used in this chapter except when appearing in direct quotes or in proper nouns like Mormon Trail. Instead, I will refer to the church as "the church" and its members as Latter-day Saints. This choice of terminology reflects the church's recent requests to use the official name rather than historic yet inaccurate names.

2. Important Latter-day Saint communities prior to their arrival in present-day Utah included Fayette and Palmyra, New York; Kirtland, Ohio; Far West, Missouri; and Nauvoo, Illinois. Recent histories have explored the relationship between the Latter-day Saints and the many Indigenous peoples living throughout the Intermountain West, noting the complex and unique relationships between these groups. For example, see Jared Farmer's chapter 2 in *On Zion's Mount* for a discussion of Mormon colonization of Utah Valley beginning in 1849, the Latter-day Saints' manifold views of their Indigenous neighbors, and the violence that erupted between groups as settlement strained local resources. Farmer describes this period as one marked by "segregation and neighborliness, disdain and respect, war and peace" (80). While much has been and will continue to be written about these relationships, my purpose in this chapter is to focus specifically on Latter-day Saint water culture.

3. The Great Basin is the major geographic region to the west of the Colorado Basin. Because the rivers within the region are inadequate to supply a growing population, Utahns have augmented these waters with the Colorado through such massive diversion projects as the Central Utah Project that transfers water out of the Green River (Colorado River Basin) and delivers it to the Wasatch Front through a complex network of tunnels, dams, and pipes.

4. The Mormon pioneers also adopted Spanish-Hispano architectural designs, as they constructed their first homes in the Salt Lake Valley out of adobe (Arrington 45). Numerous historians have addressed the false notion that the pioneers found a desert wasteland when they arrived in the Salt Lake Valley. While more arid than their previous homes farther east, the Latter-day Saints entered a valley with ample grasses and water for crops and livestock. See Tom Alexander's "Irrigating the Mormon Heartland," Jared Farmer's *On Zion's Mount*, and J. Michael Hunter's "The Desert Shall Blossom as a Rose."

5. In her article "The 'Lion of the Lord' and the Land," Sara Dant traces Young's supposed quote, as no primary textual evidence exists according to her research. Rather, in her opinion, Arrington—from whom all subsequent scholars quote Young's phrase—"simply synthesized Young's basic ideas into perfectly pithy prose" (32).

6. Prior to Powell's expedition, Catholic missionary Father Pierre-Jean De Smet in 1851 wrote of the "success of the Mormons, who in less than five years have changed the face of a frightful desert and live there in great abundance" (qtd. in Arrington and May 6).

7. In this report, Powell divides Utah into two hydrographic regions: one drained by the Colorado River and the other comprising the desert basins that have no outlet, which would constitute today's Great Basin.

8. Arrington's conclusion in *Great Basin Kingdom* reads: "As the waste involved in the short-sighted, unplanned, and ruthless exploitation of other Western frontiers became more apparent, the Mormon pattern became increasingly appreciated—became recognized as prophetic of the pattern which the entire West would ultimately have to adopt.... The design of the Kingdom, once despised and backward, is now part of the heritage which Americans are passing on to governments and people in many parts of the world" (411–12).

9. Allotments of ten to twenty acres were typical for farmers, and these parcels were held farther out of town than the smaller plots given to those living and working in the city.

10. Gamaliel was a respected Pharisee at the time of Jesus by whom the apostle Paul was trained before his conversion to Christianity. See Acts 5:34 and 22:3 in the New Testament.

11. McClurg was a resident of Colorado Springs, Colorado, and was the first woman to visit the Mesa Verde Cliff Dwellings. She was integral in the preservation of these ruins.

12. Oliver B. Huntington and John McEwan served as mission clerks, and it is to them the journal is credited. The August 21 entry, written in different penmanship from the preceding pages, notes that Huntington had returned west to see his family. As McEwan gives his name at the end of the journal, one assumes he finished the account.

13. Peterson explains that Brigham Young had conceived of this idea through his discussions with Thomas L. Kane, a US Army general and friend to the Latter-day Saints, who had envisioned a string of colonies far into Mexico that would later open up economic routes supported by sea ports, much like what Young sought to accomplish through the colonies stretching southwest toward Southern California (6).

14. Brigham Young had urged the pioneers not to pursue mining ventures but to place agriculture at the forefront of their activities when they arrived in the valley. By the late nineteenth century, however, mining become a significant enterprise in Utah (Alexander, "Stewardship and Enterprise").

15. Utah senator Jacob Anderegg introduced SB 52.

16. See Alexander's footnote 14 on page 50 for an extensive list of publications between the 1970s and 1990s that reflect this remembering.

17. Some of the publications include Jedediah S. Rogers and Matthew C. Godfrey's *The Earth Will Appear as the Garden of Eden: Essays on Mormon Environmental History* (2019) and George Handley's *The Hope of Nature: Our Care for God's Creation* (2020).

18. See Gen. 4:1–9.

19. According to the 2020 Yale Climate Opinion Maps, 67 percent of Utahns believe in climate change. Significant variation exists throughout the state, as 75 percent of those living in Salt Lake accept climate change, while 57 percent of those in the more rural Vernal do not. States throughout the Midwest and northern Rockies exhibit slightly higher denial rates. See Jennifer Marlon et al.'s interactive map for more information.

20. Elder Nash first presented these remarks at the annual Stegner Center Symposium

at the University of Utah in 2013. His remarks are one of the many links on the Church of Jesus Christ of Latter-day Saints' Environmental Stewardship and Conservation web page.

21. Founded in 1842, the Relief Society is the church's women's organization. Eubank is also the director of LDS Charities, which coordinates the church's humanitarian aid throughout the world.

22. In the church's October 2020 General Conference, Elder Gong used the phrase once again to speak of a Native family in the Four Corners region who "grows a beautiful rose bush to blossom in the desert, symbolic of gospel faith and self-reliance" ("All Nations, Kindreds, and Tongues").

Works Cited

Abbey, Edward. "The Damnation of a Canyon." *The Glen Canyon Reader*, edited by Mathew Barrett Gross, U of Arizona P, 2003, pp. 130–36.

———. *Desert Solitaire: A Season in the Wilderness*. 1968. Ballantine, 1971.

"About." Nuestro Río, Facebook, www.facebook.com/pg/NuestroRio/about/?ref=page_internal.

"About the Weekly Lawn Watering Guide." *Division of Water Resources Conservation Program*, Utah Division of Water Resources, conservewater.utah.gov/guide.html.

Abruzzi, William. *Dam That River! Ecology and Mormon Settlement in the Little Colorado River Basin*. UP of America, 1993.

"Accumulations for October 2021." Lower Colorado River Daily Report: Current Month, Lower Colorado Region, Bureau of Reclamation, 7 Oct. 2021. www.usbr.gov/lc/region/g4000/hourly/levels.html.

Adamson, Joni, Mei Mei Evans, and Rachel Stein, editors. "Environmental Justice Politics, Poetics, and Pedagogy." Introduction. *The Environmental Justice Reader*, U of Arizona P, 2002, pp. 3–14.

Adamson, Joni, and Salma Monani. "Cosmovisions, Ecocriticism, and Indigenous Studies." Introduction. *Ecocriticism and Indigenous Studies: Conversations from Earth to Cosmos*, edited by Salma Monani and Joni Adamson, Routledge, 2017, pp. 1–19.

Adler, Robert W. *Restoring Colorado River Ecosystems: A Troubled Sense of Immensity*. Island Press, 2007.

Ahasteen, Jack. "Control the Water and You Have Everything." Cartoon. *Navajo Times*, 21 May 2020, navajotimes.com/wp-content/uploads/2020/05/Jacks-5-21-20.gif.

———. "$600 Million Care Act Water Line." Cartoon. *Navajo Times*, 28 May 2020, navajotimes.com/wp-content/uploads/2020/05/WATERLINE5282020.gif.

Alexander, Thomas. "Irrigating the Mormon Heartland: The Operation of the Irrigation Companies in Wasatch Oasis Communities, 1847–1880." *Agricultural History*, vol. 76, no. 2, Spring 2002, pp. 172–87. JSTOR, https://www.jstor.org/stable/3744997.

———. "Lost Memory and Environmentalism: Mormons on the Wasatch Front, 1847–1930." *The Earth Will Appear as the Garden of Eden: Essays on Mormon Environmental History*, edited by Jedediah S. Rogers and Matthew C. Godfrey, U of Utah P, 2019, pp. 47–68.

———. "Stewardship and Enterprise: The LDS Church and the Wasatch Oasis Environment, 1847–1930." *Stewardship and the Creation: LDS Perspectives on the Environment*, edited by George B. Handley, Terry B. Ball, and Steven L. Peck, Religious Studies Center, 2006, pp. 15–32. Religious Studies Center, rsc.byu.edu/stewardship-creation/stewardship-enterprise.

Allison, Steven D., and Tyrus Miller. "Why Science Needs the Humanities to Solve Climate Change." *The Conversation*, 1 Aug. 2019, theconversation.com/why-science-needs-the-humanities-to-solve-climate-change-113832.

Alvarez, Clemente, and Nacho Corbella. *Colorado River: Hispanics to the Rescue*, Univision Planeta, Univision Communications, 2015, huelladigital.univisionnoticias.com/riocolorado/1-a-river-on-the-border/?lang=en.

Anderson, D. H. Editorial. *The Irrigation Age*, vol. 33, no. 1, Mar. 1918, pp. 5–7. L. Tom Perry Special Collections, Brigham Young University, Provo, Utah.

Anderson, D. Larry. "Utah's Perspective: The Colorado River." *Utah Division of Water Resources*, 2nd ed., May 2002, water.utah.gov/wp-content/uploads/2019/01/The ColoradoRiverart.pdf.

Arellano, Juan Estevan. "La Querencia: La Raza Bioregionalism." *New Mexico Historical Review*, vol. 72, no. 1, 1997, pp. 31–37.

"Arizona v. California U.S. Supreme Court Decision of 1964." *Law of the River*, Bureau of Reclamation, www.usbr.gov/lc/region/pao/lawofrvr.html#avc.

Arlen, Caroline. "Journey to the bottom of Navajo Lake." *High Country News*, 4 Aug. 2003, www.hcn.org/issues/255/14154.

Armitage, Susan H. "Through Women's Eyes: A New View of the West." *The Women's West*, edited by Susan H. Armitage and Elizabeth Jameson, U of Oklahoma P, 1987, pp. 9–18.

Arrington, Leonard J. *Great Basin Kingdom: An Economic History of the Latter-day Saints, 1830–1900*. Harvard UP, 1958.

Arrington, Leonard, and Dean May. "'A Different Mode of Life': Irrigation and Society in Nineteenth-Century Utah." *Agricultural History*, vol. 49, no. 1, Jan. 1975, pp. 3–20. JSTOR, www.jstor.org/stable/3742105.

Asociación Ecológica de Usuarios del Río Hardy-Colorado, A.C. (AEURHYC). "Plan Estratégico." *Colorado River Delta/Delta del Rio Colorado Research Coordination Network*, University of Arizona, 2004, www.geo.arizona.edu/rcncrd/documents/Plano_Estrategico.pdf.

Austin, Mary. *The Land of Little Rain*. 1903. Modern Library, 2003.

Babbitt, Bruce. "Here's How Less than 10% of Farmland Could Solve the Colorado River's Water Deficit." *AZ Central*, 12 May 2020, www.azcentral.com/story/opinion/op-ed/2020/05/12/colorado-river-overdrawn-retire-farmland-can-solve/3109406001/.

Bacigalupi, Paolo. *The Water Knife*. Vintage, 2015.

Barlow, Maude, and Tony Clarke. *Blue Gold: The Fight to Stop the Corporate Theft of the World's Water*. New Press, 2002.

Bauman, Whitney A., Richard R. Bohannon II, and Kevin J. O'Brien, editors. Introduction. *Grounding Religion: A Field Guide to the Study of Religion and Ecology*, Routledge, 2011, pp. 1–9.

Baym, Nina. *Women Writers of the American West, 1833–1927*. U of Illinois P, 2001. JSTOR, doi-org.usd.idm.oclc.org/10.5406/j.ctt1xcjg0.

Beale, Walter H. *A Pragmatic Theory of Rhetoric*. Southern Illinois UP, 1987.

Beard, Dan. *Deadbeat Dams: Why We Should Abolish the U.S. Bureau of Reclamation and Tear Down Glen Canyon Dam*. Johnson, 2015.

Belnap, Jayne. "Cryptobiotic Soils: Holding the Place in Place." *Impacts of Climate Changes on Life and Ecosystems*, United States Geological Survey, 18 May 2004, geochange.er.usgs.gov/sw/impacts/biology/crypto/.

Bennett, Jane. *Vibrant Matter: A Political Ecology of Things*. Duke UP, 2010.

Bennion, John. *Ezekiel's Third Wife*. Roundfire Books, 2019.

———. "Water Law on the Eve of Statehood: Israel Bennion and a Conflict in Vernon, 1893–1896." *Utah Historical Quarterly*, vol. 82, no. 4, 2014, pp. 289–305. *Issuu*, issuu.com/utah10/docs/uhq_volume82_2014_number4.

Berg, Peter, and Raymond Dasmann. "Reinhabiting California." *The Biosphere and the Bioregion: Essential Writings of Peter Berg*, edited by Cheryl Glotfelty and Eve Quensel, Taylor & Francis, 2014, pp. 35–40. Google Books, hwww.google.com/books/edition/The_Biosphere_and_the_Bioregion/dF4KBAAAQBAJ?hl=en&gbpv=0.

Bergman, Charles. *Red Desert: Fighting for Life at the End of the Colorado River*. Fulcrum, 2002.

Bergthaller, Hannes, et al. "Mapping Common Ground: Ecocriticism, Environmental History, and the Environmental Humanities." *Environmental Humanities*, vol. 5, 2014, pp. 261–76, Duke UP, doi.org/10.1215/22011919-3615505.

Blanchard, C. J. *The Mentor: Reclaiming the Desert*, vol. 6, no. 17, serial 165, 15 Oct. 1918. Project Gutenberg, www.gutenberg.org/files/49454/49454-0.txt.

The Book of Mormon: Another Testament of Jesus Christ. The Church of Jesus Christ of Latter-day Saints, 2005.

"Boulder Canyon Project Act." *Law of the River*, Lower Colorado Region, Bureau of Reclamation, www.usbr.gov/lc/region/pao/lawofrvr.html#bcp.

Bounds, Susanne, and Patti Capel Swartz. "Women's Resistance in the Desert West." *Literature of Nature: An International Sourcebook*, edited by Patrick Murphy and Fitzroy Dearborn, 1998, pp. 77–83.

Brean, Henry. "Heavy Winter Fall Will Likely Prevent Lake Mead Shortage for 2018." *Boulder City Review*, 1 Mar. 2017, bouldercityreview.com/news/heavy-winter-fall-will-likely-prevent-lake-mead-shortage-2018.

Brooks, Cleanth. *The Well Wrought Urn*. Reynal & Hitchcock, 1947.

Brown, Bill. "Reading the West: Cultural and Historical Background." *Reading the West: An Anthology of Dime Westerns*. Bedford/St. Martin's, 1997, pp. 1–40.

Buell, Lawrence. *The Environmental Imagination: Thoreau, Nature Writing, and the Formation of American Culture*. Harvard UP, 1995.

———. *Writing for an Endangered World: Literature, Culture, and Environment in the U.S. and Beyond*. Belknap Press, 2001.

Bullis, Connie. "Retalking Environmental Discourses from a Feminist Perspective: The Radical Potential of Ecofeminism." *The Symbolic Earth: Discourse and Our Creation of the Environment*, edited by James G. Cantrill and Christine L. Oravec. UP of Kentucky, 1996, pp. 123–48.

Buyer, Laurie Wagner. *Side Canyons*. Five Star, 2004.

Campbell, Neil. *The Cultures of the American New West*. Edinburgh UP, 2000.

Carpe Diem West. "Thinking Like a River Basin: Leaders' Perspectives on Options and Opportunities in Colorado River Management." *Center for Natural Resources &*

Environmental Policy, University of Montana, 2011. www.carpediemwest.org/wp-content/uploads/Thinking_Like_A_River_Basin_8-20-13.pdf.

Casey, T. Timothy. "Not in Our Lands: A Canadian Comparative Case Study of Indigenous Resistance Strategies to Natural Resource Development in British Columbia and the Arctic." *Indigenous Environmental Justice*, edited by Karen Jarratt-Snider and Marianne O. Nielsen, U of Arizona P, 2020, pp. 160–78. ProQuest, ebookcentral.proquest.com/lib/usd/detail.action?docID=6182068.

Cech, Thomas V. *Principles of Water Resources: History, Development, Management, and Policy*. John Wiley & Sons, 2003.

Childs, Craig L. *The Desert Cries: A Season of Flash Floods in a Dry Land*. Arizona Highways, 2002.

———. "The Memory of Water: Life in Ephemeral Water Holes." *House of Rain.com*, www.houseofrain.com/pdfs/ChildsMemoryofwaterWingsSpr09.pdf.

———. *The Secret Knowledge of Water: Discovering the Essence of the American Desert*. Little, Brown, 2000.

———. *The Way Out: A True Story of Ruin and Survival*. Little, Brown, 2004.

Clark, Georgie W. "Conquering Glen Canyon." *A Colorado River Reader*, edited by Richard F. Fleck, U of Utah P, 2000, pp. 97–111.

Clark, Georgie White, and Duane Newcomb. *Georgie Clark: Thirty Years of River Running*. Chronicle, 1977.

Clifton, James. *Before Obliteration by the Navajo Reservoir: Celebration of Santa Rosa Day, July 24, 1960*. Box 2, Folder 31. Frances Léon Quintana Papers. Center for Southwest Research, University of New Mexico, Albuquerque.

"The Colorado River." Wikipedia, 11 Apr. 2012, en.wikipedia.org/wiki/Colorado_River.

Colorado River Basin Ten Tribes Partnership Tribal Water Study: Study Report. Colorado River Basin Ten Tribes Partnership, Bureau of Reclamation, Dec. 2018, www.usbr.gov/lc/region/programs/crbstudy/tws/finalreport.html.

"Colorado River Basin Tribes." *Tribal Water Uses in the Colorado River Basin*. Tribalwateruse, www.tribalwateruse.org.

"Colorado River Compact of 1922." *Law of the River*, Lower Colorado Region, Bureau of Reclamation, www.usbr.gov/lc/region/pao/lawofrvr.html#compact.

Colorado River Delta/Delta Del Rio Colorado Research Coordination Network. Department of Geosciences, University of Arizona, www.geo.arizona.edu/rcncrd/index.html.

"Colorado River Storage Project Act of 1956." *Law of the River*, Lower Colorado Region, Bureau of Reclamation, www.usbr.gov/lc/region/pao/lawofrvr.html#crsp.

Comer, Krista. *Landscapes of the New West: Gender and Geography in Contemporary Women's Writing*. U of North Carolina P, 1999.

"Comments of Ival V. Goslin, Engineer-Secretary of Upper Colorado River Commission to Hon. Henry Aldous Dixon of Utah on Dec. 26, 1957." Arthur Watkins Papers. Box 29, Folder 15. L. Tom Perry Special Collections, Brigham Young University, Provo, Utah.

Cook-Lynn, Elizabeth. "Assigning Meaning to the Lewis and Clark Adventure and the Western Story." *This Stretch of the River: Lakota, Dakota, and Nakota Responses to the Lewis and Clark Expedition and Bicentennial*, edited by Craig Howe and Kim Tall Bear, Pine Hill Press, 2006, pp. 63–67.

Crowley, Sharon. *Toward a Civil Discourse: Rhetoric and Fundamentalism*. U of Pittsburgh P, 2006.

Dant, Sara. "The 'Lion of the Lord' and the Land." *The Earth Will Appear as the Garden of Eden: Essays on Mormon Environmental History*, edited by Jedediah S. Rogers and Matthew C. Godfrey, U of Utah P, 2019, pp. 29–46.

Davis, Wade. *River Notes: A Natural and Human History of the Colorado*. Island Press, 2013.

Death of a River: The Colorado River Delta. Executive producers Alexandra Cousteau and Jonathan Smith, Blue Legacy, 2010. *Vimeo*, vimeo.com/15096754.

Deems, J. S., et al. "Combined Impacts of Current and Future Dust Deposition and Regional Warming on Colorado River Basin Snow Dynamics and Hydrology." *Hydrology and Earth Systems Sciences*, vol. 17, 2013, pp. 4401–13. Copernicus Publications, doi.org/10.5194/hess-17-4401-2013.

Dellenbaugh, Frederick S. *A Canyon Voyage: The Narrative of the Second Powell Expedition*. 1908. U of Arizona P, 1996.

———. *The Romance of the Colorado River: The Story of Its Discovery in 1540, with an Account of the Later Expeditions, and with Special Reference to the Voyages of Powell Through the Line of the Great Canyons*. Putnam, 1903.

DeLoughery, Elizabeth, Jill Didur, and Anthony Carrigan. "A Postcolonial Environmental Humanities." Introduction. *Global Ecologies and the Environmental Humanities: Postcolonial Approaches*, Routledge, 2015, pp. 1–32.

Diaz, Natalie. *Postcolonial Love Poem*. Graywolf Press, 2020.

———. *When My Brother Was an Aztec*. Copper Canyon Press, 2012.

Dillion, Grace L. *"Walking the Clouds": An Anthology of Indigenous Science Fiction*. U of Arizona P, 2012.

The Doctrine and Covenants. The Church of Jesus Christ of Latter-day Saints, 2005.

Dominy, Floyd. "Lake Powell: Jewell of the Colorado." *The Glen Canyon Reader*, edited by Mathew Barrett Gross, U of Arizona P, 2003, pp. 126–28.

Draft Report: Review of the Colorado River Interim Guidelines for Lower Basin Shortages and Coordinated Operations for Lake Powell and Lake Mead. Bureau of Reclamation, Oct. 2020, www.usbr.gov/lc/region/programs/strategies/docs/7D_Kickoff_Webinars_March2020_Final_508.pdf.

Drought Contingency Plans Basin States Transmittal Letter and Attachments. Bureau of Reclamation, 19 Mar. 2019, www.usbr.gov/dcp/index.html.

"Drought in the Colorado River Basin." Department of the Interior, www.doi.gov/water/owdi.cr.drought/en/.

Dutton, C. E. "Irrigable Lands of the Valley of the Sevier River." *Report on the Lands of the Arid Region of the United States with a More Detailed Account of the Lands of Utah*. 1878. 2nd ed., edited by J. W. Powell, Government Printing Office, 1879, pp. 128–49.

Elk Mountain Mission Journal, 1855 May–October. MS 2204, Church History Library, Church of Jesus Christ of Latter-day Saints, Salt Lake City, Utah, 4 Mar. 2020, catalog.churchofjesuschristorgassets?id=40510d50-92ac-489a-b106-d098cb9f6b7a&crate=&index=2.

Ellis, Erle C. "Science Alone Won't Save the Earth. People Have to Do That." *The New York Times*, 11 Aug. 2018, www.nytimes.com/2018/08/11/opinion/sunday/science-people-environment-earth.html.

Emmett, Robert S., and David E. Nye. *The Environmental Humanities: A Critical Introduction*. MIT Press, 2017.

"Environmental Justice." United States Environmental Protection Agency, updated 8 May 2020, www.epa.gov/environmentaljustice.

"Environmental Stewardship and Conservation." Newsroom, Church of Jesus Christ of Latter-day Saints, newsroom.churchofjesuschrist.org/article/environmental-stewardship-conservation.

Estes, Nick. *Our History Is the Future: Standing Rock Versus the Dakota Access Pipeline, and the Long Tradition of Indigenous Resistance*. Verso, 2019.

Eubank, Sharon. "That We May Be One." *LDS Earth Stewardship*, ldsearthstewardship.org/component/k2/2870?Itemid=101.

Ewert, Eric C. "The Coming Challenge." *Desert Water: The Future of Utah's Water Resources*, edited by Hal Crimmel, U of Utah P, 2014, pp. 13–28.

Farmer, Jared. *Glen Canyon Dammed: Inventing Lake Powell and the Canyon Country*. U of Arizona P, 1999.

———. *On Zion's Mount: Mormons, Indians, and the American Landscape*. Harvard UP, 2008.

Fedarko, Kevin. *The Emerald Mile: The Epic Story of the Fastest Ride in History Through the Heart of the Grand Canyon*. Scribner, 2013.

Ferril, Thomas Hornsby. "Here Is a Land Where Life Is Written in Water." *Thomas Hornsby Ferril and the American West*, edited by Robert C. Baron, Stephen J. Leonard, and Thomas J. Noel, Fulcrum, 1996, pp. 36–37.

Fiege, Mark. *Irrigated Eden: The Making of an Agricultural Landscape in the American West*. U of Washington P, 1999.

Fisher, Kristina Gray. "Reclaiming Querencia: The Quest for Culturally Appropriate, Environmentally Sustainable Economic Development in Northern New Mexico." *Natural Resources Journal*, vol. 48, no. 2, 2008, pp. 479–532. University of New Mexico Digital Repository, digitalrepository.unm.edu/nrj/vol48/iss2/10/.

Fitzgerald, Stephanie. *Native Women and Land: Narratives of Dispossession and Resurgence*. U of New Mexico P, 2015.

Fleck, John. *Water Is for Fighting over and Other Myths About Water in the West*. Island Press, 2016.

Fletcher, Colin. *River: One Man's Journey Down the Colorado, Source to Sea*. Knopf, 1997.

Fonseca-Chávez, Vanessa, Levi Romero, and Spencer R. Herrera, editors. *Querencia: Reflections on the New Mexico Homeland*. U of New Mexico P, 2020.

Formisano, Paul. "Shifting Tides: A Literary Exploration of the Colorado River Delta." *Make Waves: Water in Contemporary Literature and Film*, edited by Paula Anca Farca, U of Nevada P, 2019, pp. 155–72.

Fradkin, Philip L. *A River No More: The Colorado River and the West*. 1968. U of California P, 1995.

Frehner, Brian. "'The People Cannot Conquer the River': Mormons and Water in the Arid Southwest, 1865–1938." *The Earth Will Appear as the Garden of Eden: Essays on Mormon Environmental History*, edited by Jedediah S. Rogers and Matthew C. Godfrey, U of Utah P, 2019, pp. 173–94.

Fulford, Tim, Debbie Lee, and Peter J. Kitson. *Literature, Science and Exploration in the Romantic Era: Bodies of Knowledge*. Cambridge UP, 2004.
Fullan, Rebecca Lynne. "Thomas King Tells a Different Story: Dams, Rivers, and Indigenous Literary Hydromythology." *Make Waves: Water in Contemporary Literature and Film*, edited by Paula Anca Farca, U of Nevada P, 2019, pp. 137–54.
Fuller, Craig. "Irrigation in Utah." *Utah History Encyclopedia*, www.uen.org/utah_history_encyclopedia/i/IRRIGATION.shtml.
Gemein, Mascha N. "'Seeds Must Be Among the Greatest Travelers of All': Native American Literatures Planting the Seeds for a Cosmopolitical Environmental Justice Discourse." *ISLE*, vol. 23, no. 3, 2016, pp. 485–505.
Gersdorf, Catrin. *The Poetics and Politics of the Desert: Landscape and the Construction of America*. Rodopi, 2009.
Getches, David H. *Water Law in a Nutshell*. 4th ed., Thomson West, 2009.
Gilio-Whitaker, Dina. *As Long as Grass Grows: The Indigenous Fight for Environmental Justice, from Colonization to Standing Rock*. Beacon Press, 2019.
Gilman, Charlotte Perkins. "The Yellow Wallpaper." 1892. *Herland and Selected Stories*, Signet Classics, 1992.
Glaser, Leah S. "Navajo Indian Irrigation Project." *Navajo Indian Irrigation Project*, Bureau of Reclamation, 1998. www.usbr.gov/projects/pdf.php?id=141.
———. "San Juan–Chama Project History." *San Juan–Chama Project*, Bureau of Reclamation, www.usbr.gov/projects/pdf.php?id=186.
"Glen Canyon Dam." Projects and Facilities, Bureau of Reclamation, www.usbr.gov/projects/index.php?id=144.
Goetzmann, William H. *Exploration and Empire: The Explorer and the Scientist in the Winning of the American West*. Texas State Historical Association, 1993.
Gong, Gerrit W. "All Nations, Kindreds, and Tongues." *Liahona*, Church of Jesus Christ of Latter-day Saints, Nov. 2020, www.churchofjesuschrist.org/study/general-conference/2020/10/24gong?lang=eng.
———. "Elder Gong Addresses 'Faith, Hope and Goodwill' at G20 Interfaith Forum in Japan." *Church News*, 8 June 2019, www.thechurchnews.com/leaders-and-ministry/2019-06-08/elder-gong-addresses-faith-hope-and-goodwill-at-g20-interfaith-forum-in-japan-1133.
Gottlieb, Roger S. *The Oxford Handbook of Religion and Ecology*. Oxford UP, 2006.
Grace, Stephen. *Dam Nation: How Water Shaped the West and Will Determine Its Future*. Globe Pequot Press, 2012.
Groover, Kristina K. *The Wilderness Within: American Women Writers and Spiritual Quest*. U of Arkansas P, 1999.
Handley, George B. *Home Waters: A Year of Recompenses on the Provo River*. U of Utah P, 2010.
———. *The Hope of Nature: Our Care for God's Creations*. BYU Neal A. Maxwell Institute, 2020.
Hansman, Heather. *Downriver: Into the Future of Water in the West*. U of Chicago P, 2019.
Harper, Kelly C. "The Mormon Role in Irrigation Beginnings and Diffusions in

the Western States: An Historical Geography." Thesis. Brigham Young U, 1974. BYU ScholarsArchive, scholarsarchive.byu.edu/etd/4764.

Heise, Ursula K. "Planet, Species, Justice—and the Stories We Tell About Them." Introduction. *The Routledge Companion to the Environmental Humanities*, edited by Ursula K. Heise, Jon Christensen, and Michelle Niemann, Routledge, 2017, pp. 1–13.

———. *Sense of Place, Sense of Planet: The Environmental Imagination of the Global*. Oxford UP, 2008.

Hemingway, Ernest. *Death in the Afternoon*. 1932. Charles Scribner's Sons, 1960.

Hinojosa-Huerta, Osvel. "Restoring the Colorado River Delta." *Southwest Hydrology*, vol. 3, no. 1, 2004, pp. 20–21. www.swhydro.arizona.edu/archive/v3_N1/feature5.pdf.

Hobbs, Greg. *Colorado Mother of Rivers*. Colorado Foundation for Water Education, 2005.

Hogan, Linda. "Plant Journey." *Writing Down the River: Into the Heart of the Grand Canyon*, photographed and produced by Kathleen Jo Ryan, Northland, 1998, pp. 63–68.

The Holy Bible. Authorized King James Version. The Church of Jesus Christ of Latter-day Saints, 2005.

Hooper, Troy. "Latinos Celebrate César Chávez Holiday with Song Calling for Colorado River Conservation." *The Colorado Independent*, Colorado News Collaborative, 13 Apr. 2012, www.coloradoindependent.com/2012/04/13/latinos-celebrate-cesar-chavez-holiday-with-song-calling-for-colorado-river-conservation/.

House, Ernest, Jr. "Enhancing Tribal Water Sovereignty." *DU Water Law Review Symposium*, University of Denver Sturm College of Law Water Law Review, 30 Mar. 2018, duwaterlawreview.com/events/symposium-2018/.

Hundley, Norris, Jr. *Dividing the Waters: A Century of Controversy Between the United States and Mexico*. U of California P, 1966.

———. *Water and the West: The Colorado River Compact and the Politics of Water in the American West*. 1975, U of California P, 2009.

Hunter, J. Michael. "The Desert Shall Blossom as the Rose: Pioneering Irrigation/John A. Widtsoe." *Pioneer*, 2003, pp. 10–14, 24–25. BYU ScholarsArchive, scholarsarchive.byu.edu/facpub/1404.

Hunter, Milton R. "The Mormons and the Colorado River." *The American Historical Review*, vol. 44, no. 3, Apr. 1939, pp. 549–55. JSTOR, doi:10.2307/1839902.

Hyde, Elizabeth. *In the Heart of the Canyon*. Knopf, 2009.

"In Current." YouTube, uploaded by Yeti, 7 June 2015, youtu.be/qkApBW1GC80.

Ingram, Helen. "Place Humanists at the Headgates." *Reopening the American West*, edited by Hal Rothman, U of Arizona P, 1998, pp. 155–67.

Iovino, Serenella. "Restoring the Imagination of Place: Narrative Reinhabitation and the Po Valley." *The Bioregional Imagination: Literature, Ecology, and Place*, edited by Tom Lynch, Cheryll Glotfelty, and Karla Armbruster, U of Georgia P, 2012, pp. 100–117.

Ives, Joseph C. *Report upon the Colorado River of the West*. Washington: Government Printing Office, 1861. Google Books, www.google.com/books/edition/Report_upon_the_Colorado_River_of_the_We/UgLWCizxCSgC?hl=en&authuser=0.

Jarratt-Snider, Karen, and Marianne O. Nielsen, editors. Introduction. *Indigenous Environmental Justice*. U of Arizona P, 2020, pp. 3–15. ProQuest, ebookcentral.proquest.com/lib/usd/detail.action?docID=6182068.

Johnson, Susan Lee. "'A Memory Sweet to Soldiers': The Significance of Gender in the History of the American West." *Western Historical Quarterly*, vol. 24, no. 4, 1993, pp. 495–517. JSTOR, doi.org/10.2307/970703.

Jones, Paul. "Congressional Record February 7, 1956." Box 29, Folder 16. MSS 146 V. Watkins Papers. L. Tom Perry Special Collections, Brigham Young University, Provo, Utah.

———. "February 8, 1956." Box 29, Folder 16. MSS 146 Arthur V. Watkins Papers. L. Tom Perry Special Collections, Brigham Young University, Provo, Utah. 16 June 2015.

———. "June 29, 2015." Nuestro Río, Facebook, www.facebook.com/230421010324096/videos/1000429516656571.

Kang, Yeonhaun. "The Garden in Motion: Botanical Exchange and Transnational Collaboration in Leslie Marmon Silko's *Gardens in the Dunes*." *Studies in American Fiction*, vol. 46, no. 1, 2019, pp. 103–29. Project Muse, doi.org/10.1353/saf.2019.0004.

Kelley, Klara, and Harris Francis. "COVID-19 Crisis Is Partly a Water Crisis." *Navajo Times*, 21 May 2020, navajotimes.com/opinion/letters/letters-covid-19-crisis-is-partly-a-water-crisis/.

Kino, Eusebio Francisco. *Plan for the Development of Pimería Alta, Arizona & Upper California, a Report to the Mexican Viceroy*. Translated and annotated by Ernest J. Burrus, Arizona Pioneers' Historical Society, 1961.

Kober, Amy Souers. "Announcing America's Most Endangered Rivers of 2017." *American Rivers*, 11 Apr. 2017, www.americanrivers.org/2017/04/most-endangered-rivers-2017/?js=false&source=adwords&gclid=CjwKEAjwl9DIBRCG_e3DWSKsizSSJADMmjI1b90cxoss3i_u-MTSMXVCslIXISWNidDDFl_8kcs3BROcdw7w_WCB.

Koch, Marilyn J. "The Indian Camp Dam Controversy: The Real Beanfield War." *The Bureau of Reclamation: History Essays from the Centennial Symposium*, vol. 2, 2008, pp. 551–77. Google Books, www.google.com/books/edition/The_Bureau_of_Reclamation/ookOiUDdfSEC?hl=en&gbpv=0.

Kolb, Ellsworth L. *Through the Grand Canyon from Wyoming to Mexico*. 1912. Macmillan, 1952.

Kollin, Susan. "Postwestern Studies, Dead or Alive." Introduction. *Postwestern Cultures: Literature, Theory, Space*, edited by Susan Kollin, U of Nebraska P, 2007, pp. ix–xix.

Kolodny, Annette. *The Lay of the Land: Metaphor as Experience and History in American Life and Letters*. U of North Carolina P, 1975.

Krech, Shepard. *The Ecological Indian: Myth and History*. Norton, 2000.

Krutch, Joseph Wood. *The Desert Year*. 1952. U of Arizona P, 1985.

———. *The Voice of the Desert: A Naturalist's Interpretation*. William Sloane, 1954.

Kuhn, Eric, and John Fleck. *Science Be Dammed: How Ignoring Inconvenient Science Drained the Colorado River*. U of Arizona P, 2019.

Lamont, Victoria. *Westerns: A Woman's History*. U of Nebraska P, 2016.

Lavender, David. *Colorado River Country*. E. P. Dutton, 1982.

———. *River Runners of the Grand Canyon*. Grand Canyon Natural History Association, 1985.

Lawson, Angelica. "Resistance and Resilience in Ofelia Zepeda's *Ocean Power*." *The Kenyon Review*, vol. 32, no. 1, 2010, pp. 180–98. JSTOR, www.jstor.com/stable/40600285.

Leche y miel. Directed by Justin Clifton, Hispanic Access Foundation and American Rivers, 2017, www.americanrivers.org/rivers/films/milk-and-honey/.

Lee, Katie. *All My Rivers Are Gone: A Journey of Discovery Through Glen Canyon*. Johnson Books, 1998.

———. "Split Bar." *Hidden Passage: The Journal of Glen Canyon Institute*, no. 11, Winter 2004, pp. 10–13.

Leich, Harold H. *Alone on the Colorado*. 1938. U of Utah P, 2019.

Lejano, Raul, Mrill Ingram, and Helen Ingram. *The Power of Narrative in Environmental Networks*. MIT Press, 2013.

Lentfer, Hank, and Carolyn Servid, editors. *Arctic Refuge: A Circle of Testimony*. Milkweed, 2001.

Leopold, Aldo. *A Sand County Almanac: With Essays on Conservation from Round River*. 1949. Ballantine, 1970.

"Level III and IV Ecoregions of the Continental U.S." *EPA*, 18 June 2020, www.epa.gov/eco-research/level-iii-and-iv-ecoregions-continental-united-states.

Limerick, Patricia. *Desert Passages: Encounters with the American Deserts*. U of New Mexico P, 1985.

———. *The Legacy of Conquest: The Unbroken Past of the American West*. Norton, 1987.

———. *Something in the Soil: Legacies and Reckonings in the New West*. Norton, 2000.

Linenberger, Toni Rae. "The Navajo Unit: Colorado River Storage Project History." 1998. Google Books, www.usbr.gov/projects/pdf.php?id=86.

Loeffler, Jack, and Celestia Loeffler, editors. *Thinking Like a Watershed: Voices from the West*. U of New Mexico P, 2012.

Loomis, Brandon. "CAP to Meet with Upper Colorado River Officials in Response to Water 'Manipulation' Charge." *Arizona Republic*, 19 Apr. 2018, azcentral.com/story/news/local/arizona-environment/2018/04/19/cap-meet-upper-colorado-river-officials-response-water-manipulation-charge/534377002/.

Lopez, Barry. *The Rediscovery of North America*. Vintage, 1992.

Lower Colorado Water Supply Report. Bureau of Reclamation, 4 Oct. 2021, www.usbr.gov/lc/region/g4000/weekly.pdf.

Luecke, Daniel F., et al. *A Delta Once More: Restoring Riparian and Wetland Habitat in the Colorado River Delta*. Environmental Defense Fund, June 1999, www.edf.org/sites/default/files/425_delta.pdf.

Lutts, Ralph H. *The Nature Fakers: Wildlife, Science, and Sentiment*. Fulcrum, 1990.

Lynch, Tom. *Xerophilia: Ecocritical Explorations in Southwestern Literature*. Texas Tech UP, 2008.

Lynch, Tom, Cheryll Glotfelty, and Karla Armbruster, editors. *The Bioregional Imagination: Literature, Ecology, and Place*, U of Georgia P, 2012.

Lynch, Tom, et al. *Thinking Continental: Writing the Planet One Place at a Time*. U of Nebraska P, 2017.

Lyon, Thomas J. Introduction. *Writing Natural History: Dialogues with Authors*, edited by Edward Lueders, U of Utah P, 1989, pp. 1–6.

———. *This Incomparable Land: A Guide to American Nature Writing*. Milkweed, 2001.

Lyons, Scott Richard. "Rhetorical Sovereignty: What Do American Indians Want from Writing?" *College Composition and Communication*, vol. 51, no. 3, Feb. 2000, pp. 447–68. JSTOR, doi:10.2307/358744.

MacMillan, Leslie. "Lose the Crust, Inherit the Wind." *The New York Times*, 18 July 2012, green.blogs.nytimes.com/2012/07/18/lose-the-crust-inherit-the-wind/.

Maffly, Brian. "Surrounding States Bash Utah's Lake Powell Pipeline Project." *The Salt Lake Tribune*, 10 Sept. 2020, sltrib.com/news/environment/2020/09/09/surrounding-states-bash/.

Markham, Lauren. "Who Keeps Buying California's Scarce Water? Saudi Arabia." *The Guardian*, 25 Mar. 2019, www.theguardian.com/us-news/2019/mar/25/california-water-drought-scarce-saudi-arabia.

Marlon, Jennifer, et al. "Yale Climate Opinion Maps 2020." *Yale Program on Climate Change Communication*, Yale School of the Environment, 2 Sept. 2020, climatecommunication.yale.edu/visualizations-data/ycom-us/.

Martin, Russell. *A Story That Stands Like a Dam: Glen Canyon and the Struggle for the Soul of the West*. U of Utah P, 1989.

McAndrews, Kristin M. *Wrangling Women: Humor and Gender in the American West*. U of Nevada P, 2006.

McCairen, Patricia. *Canyon Solitude: A Woman's Solo Journey Through Grand Canyon*. Seal Press, 1998.

McCool, Daniel. "A New Water Ethic." *Desert Water: The Future of Utah's Water Resources*, edited by Hal Crimmel, U of Utah P, 2014, pp. 219–34.

McFarland, Sarah E. "Wild Women: Literary Explorations of American Landscapes." *Women Writing Nature: A Feminist View*, edited by Barbara J. Cook, Lexington Books, 2008, pp. 41–55.

McMillin, T. S. *The Meaning of Rivers: Flow and Reflection in American Literature*. U of Iowa P, 2010.

McPhee, John. *Encounters with the Archdruid: Narratives About a Conservationist and Three of His Natural Enemies*. Farrar, Strauss & Giroux, 1971.

Mead, Elwood. *Irrigation Institutions: A Discussion of the Economic and Legal Questions Created by the Growth of Irrigated Agricultural in the West*. 1903. Macmillan, 1910. Google Books, www.google.com/books/edition/Irrigation_Institutions/kRorAQAAIAAJ?hl=en&authuser=0.

Meloy, Ellen. *Raven's Exile: A Season on the Green River*. Henry Holt, 1994.

"The Mexican Water Treaty of 1944." *Law of the River*, Lower Colorado Region, Bureau of Reclamation, www.usbr.gov/lc/region/pao/lawofrvr.html#mexico.

Milligan, Mark. "Glad You Asked: Does Utah Really Use More Water than Any Other State?" *Survey Notes*, vol. 50, no. 2, May 2018, pp. 11–12, https://geology.utah.gov/map-pub/survey-notes/glad-you-asked/does-utah-use-more-water/.

"Minute 242 of the U.S.-Mexico International Boundary and Water Commission of 1973." *Law of the River*, Lower Colorado Region, Bureau of Reclamation, www.usbr.gov/lc/region/pao/lawofrvr.html#242.

"Minute 306." *Minutes Between the United States and Mexican Sections of the IBWC*, International Boundary and Water Commission, www.ibwc.gov/Files/Minutes/Min306.pdf.

"Minute 319." *Minutes Between the United States and Mexican Sections of the IBWC*, International Boundary and Water Commission. www.ibwc.gov/Files/Minutes/Minute_319.pdf.

"Minute 319 Colorado River Limotrophe and Delta Environmental Flows Monitoring Final Report." *Colorado River Delta Pulse Flow*, University of Arizona, pulseflow .arizona.edu/minute-319-reports.

"Minute 323." *Minutes Between the United States and Mexican Sections of the IBWC*, International Boundary and Water Commission, www.ibwc.gov/Files/Minutes /Min323.pdf.

Montoya, Joseph M. "Letter to President Nixon." Box 149, Folder 10. Folder, MSS 386 BC Joseph M. Montoya Papers, Center for Southwest Research, University of New Mexico, Albuquerque.

———. "Representative Montoya to Subcommittee on Irrigation and Reclamation, House Committee on Interior and Insular Affairs, May 20, 1960." Box 65, Folder 30. MSS 386 BC Joseph M. Montoya Papers, Center for Southwest Research, University of New Mexico, Albuquerque.

———. "Statement 1962." Box 65, Folder 30. MSS 386 BC Joseph M. Montoya Papers, Center for Southwest Research, University of New Mexico, Albuquerque. 10 June 2016.

Moore, Bryan L. *Ecology and Literature: Ecocentric Personification from Antiquity to the Twenty-First Century*. Palgrave Macmillan, 2008.

Moore, Roberta, and Scott Slovic. *Wild Nevada: Testimonies on Behalf of the Desert*. U of Nevada P, 2005.

Mote, Philip W., et al. "Dramatic Declines in Snowpack in the Western U.S." *NPJ Climate and Atmospheric Science*, vol. 1, no. 2, 2018, pp. 1–6. *Nature*, doi:10.1038/s41612 -018-0012-1.

Nash, Marcus B. "Righteous Dominion and Compassion for the Earth." *The Earth Will Appear as the Garden of Eden: Essays on Mormon Environmental History*, edited by Jedediah S. Rogers and Matthew C. Godfrey, U of Utah P, 2019, pp. 263–74.

Nash, Roderick Frazier. *Wilderness and the American Mind*. 4th ed., Yale UP, 2001.

"Navajo Dam." Projects and Facilities, Bureau of Reclamation, www.usbr.gov/projects /index.php?id=310.

Navajo Dam Dedication Program. 15 Sept. 1962. Box 31, Folder 22. MSS 20 BC Clinton P. Anderson Papers, Center for Southwest Research, University of New Mexico, Albuquerque.

"Navajo-Gallup Water Supply Project." *Interior Region 7—Upper Colorado Basin*. Bureau of Reclamation. Last updated 1 July 2021. www.usbr.gov/uc/progact/navajo-gallup /index.html.

"Navajo Unit." *Upper Colorado Region*, Bureau of Reclamation, www.usbr.gov/uc/rm/crsp /navajo/index.html.

Nelson, Cody. "'This Land Is All We Have Left': Tribes on Edge over Giant Dam Proposal near Grand Canyon." *The Guardian*, 12 Aug. 2020, www.theguardian.com /us-news/2020/aug/12/navajo-nation-dams-big-canyon-pumped-storage-project.

Nelson, Russell M. "The Creation." General Conference, Church of Jesus Christ of Latter-day Saints, Apr. 2000, www.churchofjesuschrist.org/study/general-conference /2000/04/the-creation?lang=eng.

"Nez Testifies on Capitol Hill in Support of Scholarship Funding." *Navajo-Hopi Observer*,

18 Feb. 2020, www.nhonews.com/news/2020/feb/18/nez-testifies-capitol-hill-support-scholarship-fun/.

Nichols, John. *The Milagro Beanfield War*. 1974. Henry Holt, 1994.

Nixon, Rob. *Slow Violence and the Environmentalism of the Poor*. Harvard UP, 2011.

North, Mary Remsen. *Down the Colorado by a Lone Girl Scout, Mary Remsen North*. Putnam, 1930.

Norwood, Vera, and Janice Monk, editors. Introduction. *The Desert Is No Lady: Southwestern Landscapes in Women's Writing and Art*. U of Arizona P, 1987, pp. 1–9.

"Nuestro Rio, the Colorado River." YouTube, uploaded by American Rivers, 14 May 2014, www.youtube.com/watch?v=xwsw78KykNY.

O'Donoghue, Amy Joi. "Utah Lawmakers Took 'Good First Step' on Environment Worries." *Desert News*, 16 Mar. 2019, www.deseret.com/2019/3/16/20668497/utah-lawmakers-took-good-first-step-on-environment-worries.

Oelschlaeger, Max. *Caring for Creation: An Ecumenical Approach to the Environmental Crisis*. Yale UP, 1994.

Ogren, Kim, and Aaron Wolf. "Understanding Incentives and Disincentives for Conflict Prevention and Mitigation: A Case Study Examination of the Bureau of Reclamation's Implementation of the Water 2025 Initiative." Bureau of Reclamation/US Department of the Interior, Sept. 2012. Transboundary Freshwater Dispute Database, Oregon State University, transboundarywaters.science.oregonstate.edu/files/Database/ResearchProjects/WesternWater/2012%20Water2025%20Case%20Study_final%20v1.pdf.

Owen, David. *Where the Water Goes: Life and Death Along the Colorado River*. Riverhead Books, 2017.

Parrish, Alan K. *John A. Widtsoe: A Biography*. Deseret Books, 2003.

The Pearl of Great Price. The Church of Jesus Christ of Latter-day Saints, 2005.

Peña, Devon G. *Mexican Americans and the Environment: Tierra y Vida*. U of Arizona P, 2005.

———. "The Scope of Latino/a Environmental Studies." *Latino Studies*, vol. 1, no. 1, 2003, pp. 47–48.

Peterson, Charles S. *Take Up Your Mission: Mormon Colonizing Along the Little Colorado River, 1870–1900*. U of Arizona P, 1973.

Phelan, James. *Narrative as Rhetoric: Technique, Audiences, Ethics, Ideology*. Ohio State UP, 1996.

Phelan, James, and Peter J. Rabinowitz. "The Approaches." Introduction. *Narrative Theory: Core Concepts & Critical Debates*, edited by David Herman et al., Ohio State UP, 2012, pp. 3–8.

Pike, Zebulon Montgomery. *Exploratory Travels Through the Western Territories of North America: Comprising a Voyage from St. Louis, on the Mississippi, to the Source of That River: And a Journey Through the Interior of Louisiana, and the North-Eastern Provinces of New Spain: Performed in the Years 1805, 1806, 1807, by Order of the Government of the United States*. 1811. W. H. Lawrence, 1889. Google Books, www.google.com/books//Exploratory_Travels_Through_the_Western/ztUBAAAAMAAJ?hl=en.

Porter, Eliot. *The Place No One Knew: Glen Canyon on the Colorado*. Sierra Club, 1963.

Powell, James Lawrence. *Dead Pool: Lake Powell, Global Warming, and the Future of Water in the West*. U of California P, 2008.

Powell, John Wesley. *The Arid Lands*. 1878. Edited by Wallace Stegner, U of Nebraska P, 2004.

———. *The Exploration of the Colorado and Its Canyons*. 1895. Dover, 1961.

Pozza, David M. *Bedrock and Paradox: The Literary Landscapes of Edward Abbey*. Peter Lang, 2006.

"Prize Ode to Irrigation." Words by Mrs. Gilbert McClurg, music by John J. McClellan. babel.hathitrust.org/cgi/pt?id=mdp.39015007864567&view=1up&seq=1.

"Providing in the Lord's Way." *Church Welfare Plan*, Church of Jesus Christ of Latter-day Saints, www.churchofjesuschrist.org/topics/welfare/the-church-welfare-pla /providing-in-the-lords-way?lang=eng#notes.

Purdy, John. "Water as Leitmotif in Silko's *Gardens in the Dunes*." *Reading Leslie Marmon Silko: Critical Perspectives Through Gardens in the Dunes*, edited by Laura Coltelli, Pisa UP, 2007, pp. 149–63.

Quintana (Swadesh), Frances Leon. "Discussion with Members of the Tri-Ethnic Program." KIUP Radio Program, 18 Aug. 1961. Box 5, Folder 2. MSS 768 BC Frances Léon Quintana Papers. Center for Southwest Research, University of New Mexico, Albuquerque.

———. Interview. Conducted by F. Swadesh, 18 July 1961. Box 5, Folder 1. MSS 768 BC Frances Léon Quintana Papers. Center for Southwest Research, University of New Mexico, Albuquerque.

———. "Land, Water, and Pueblo: Hispanic Relations in Northern New Mexico." Box 1, Folder 12. MSS 768 BC Frances Léon Quintana Papers. Center for Southwest Research, University of New Mexico, Albuquerque.

———. *Los Primeros Pobladores: Hispanic Americans of the Ute Frontier*. U of Notre Dame, 1974.

———. "Western History Association Address." 18 Oct. 1991. Box 1, Folder 8. MSS 768 BC Frances Léon Quintana Papers. Center for Southwest Research, University of New Mexico, Albuquerque.

Rails to Riches. Exhibit. Navajo State Park Visitor Center, Arboles, Colorado.

Rajagopalan, Balaji, et al. "Water Supply Risk on the Colorado River: Can Management Mitigate?" *Water Resources Research* 45, no. WO8201, 2009, doi: 10.1029/2008WR007652.

Ratcliffe, Krista. *Rhetorical Listening: Identification, Gender, Whiteness*. Southern Illinois UP, 2005.

"Reclamation." *Oxford English Dictionary*. www-oed-com.ezproxy.usd.edu/view/Entry /159603?redirectedFrom=reclamation#eid.

Record of Decision: Colorado River Interim Guidelines for Lower Basin Shortages and the Coordinated Operations for Lake Powell and Lake Mead. Bureau of Reclamation. Department of the Interior. www.usbr.gov/lc/region/programs/strategies/Record ofDecision.pdf.

Reisner, Marc. *Cadillac Desert: The American West and Its Disappearing Water*. Viking, 1986.

Renewal: A Reborn Colorado River Once Again Finds Her Path to the Sea. Produced by Andrew Quinn and Owen Bissell, Raise the River Coalition, 2014, sonoraninstitute .org/card/colorado-river-delta/.

Richard, Holly Jean, and Paul Formisano. *Reading Aridity in Western American Literature*, edited by Jada Ach and Gary Reger, Lexington P, 2020, pp. 141–68.

Rodríguez, Sylvia. *Acequia: Water Sharing, Sanctity, and Place*. School for Advanced Research Press, 2006.

Romero, Tom. "Welcome Address." *DU Water Law Review Symposium*, University of Denver Sturm College of Law Water Law Review, 30 Mar. 2018. YouTube, youtu.be/wXx8j8qK8tw.

Roosevelt, Franklin D. "Address at the Dedication of Boulder Dam." American Presidency Project, UC Santa Barbara, www.presidency.ucsb.edu/documents/address-the-dedication-boulder-dam.

Roosevelt, Theodore. "Address of President Roosevelt at the Tabernacle, Salt Lake City, Utah, May 29, 1903." Theodore Roosevelt Papers at the Library of Congress, Library of Congress, Dickinson State University, www.theodorerooseveltcenter.org/Research/Digital-Library/Record.aspx?libID=o289941.

Ross, Derek G. "Dam Visuals: The Changing Visual Argument for the Glen Canyon Dam." *Journal of Technical Writing and Communication*, vol. 38, no. 1, 2008, pp. 75–94. SAGE journals, doi.org/10.2190%2FTW.38.1.e.

Rowley, William D. *The Bureau of Reclamation: Origins and Growth to 1945, Volume 1*. Bureau of Reclamation, 2006. www.usbr.gov/history/.

Runyon, Luke. "How the 'Grand' Became the 'Colorado' and What It Says About Our Relationship to Nature." KUNC, 20 Dec. 2017, www.kunc.org/community/2017-12-20/how-the-grand-became-the-colorado-and-what-it-says-about-our-relationship-to-nature.

Ryan, Kathleen Jo, photographer and producer. *Writing Down the River: Into the Heart of the Grand Canyon*. Northland, 1998.

Sadler, Christa. *There's This River: Grand Canyon Boatman Stories*. 2nd ed., This Earth Press, 2006.

"Secretary Norton Signs Water Rights Agreement for Tohono O'odham Nation." *News U.S. Department of the Interior*, Department of the Interior, 30 Mar. 2006, www.doi.gov/sites/doi.gov/files/archive/news/archive/06_News_Releases/060330a.htm.

Silko, Leslie Marmon. *Gardens in the Dunes*. Simon and Schuster, 1999.

"Sister Aburto Reaches Out to Navajo Latter-day Saints in the American Southwest." Newsroom, Church of Jesus Christ of Latter-day Saints, 12 Nov. 2018, newsroom.churchofjesuschrist.org/article/sister-aburto-reaches-out-navajo-latter-day-saints-american-southwest.

Slotkin, Richard. *Gunfighter Nation: The Myth of the Frontier in Twentieth-Century America*. U of Oklahoma P, 1992.

Slovic, Scott. "Epistemology and Politics in American Nature Writing: Embedded Rhetoric and Discrete Rhetoric." *Green Culture: Environmental Rhetoric in Contemporary America*, edited by Carl G. Herndl and Stuart C. Brown, U of Wisconsin P, 1996, pp. 82–110.

———. *Seeking Awareness in American Nature Writing: Henry David Thoreau, Annie Dillard, Edward Abbey, Wendell Berry, Barry Lopez*. U of Utah P, 1992.

Smith, Henry Nash. *Virgin Land: The American West as Symbol and Myth*. 1950. Harvard UP, 1978.

Smith, Melvin T. "Mormon Exploration in the Lower Colorado River Area." *The Mormon Role in the Settlement of the West*, edited by Richard H. Jackson, Brigham Young UP, 1978, pp. 29–49.

Smythe, William E. *The Conquest of Arid America*. 1899. U of Washington P, 1970.

———. "Mormon Land Tenure." *Current Literature*, vol. 29, Oct. 1900. L. Tom Perry Special Collections, Brigham Young University, Provo, Utah.

———. "The Republic of Irrigation." *The Irrigation Age*, vol. 6, no. 5, May 1894, pp. 189–95. L. Tom Perry Special Collections, Brigham Young University, Provo, Utah.

Snyder, Gary. *The Practice of the Wild: Essays by Gary Snyder*. 1990. North Point Press, 2001.

Solnit, Rebecca. "Grounds for Hope." *Tikkun*, vol. 32, no. 1, 2017, pp. 31–39. Duke UP, doi.org/10.1215/08879982-3769066.

Sörlin, Sverker. "Environmental Humanities: Why Should Biologists Interested in the Environment Take the Humanities Seriously?" *BioScience*, vol. 62, no. 9, 2012, pp. 788–89. JSTOR, doi:10.1525/bio.2012.62.9.2.

Spinuzzi, Clay, and Mark Zachry. "Genre Ecologies: An Open-System Approach to Understanding and Constructing Documentation." *ACM Journal of Computer Documentation*, vol. 24, no. 3, 2000, pp. 169–81. *ACM Digital Library*, doi.org/10.1145/344599.344646.

Stegner, Wallace. *Beyond the Hundredth Meridian: John Wesley Powell and the Second Opening of the West*. 1954. Penguin, 1992.

———. *Mormon Country*. 1942. U of Nebraska P, 2003.

———. "Thoughts in a Dry Land." *Where the Bluebird Sings to the Lemonade Springs: Living and Writing in the West*. Penguin, 1993, pp. 45–56.

Stewart, Frank. *A Natural History of Nature Writing*. Island Press, 1995.

"Stopping Grand Canyon Escalade." Grand Canyon Trust, 2020, www.grandcanyontrust.org/stopping-grand-canyon-escalade.

Summitt, April R. *Contested Waters: An Environmental History of the Colorado River*. U of Colorado P, 2013.

Sweet, Phoebe. "Cash for Grass Program Taking Steps to Entice More Businesses." *Las Vegas Sun*, 17 June 2008, lasvegassun.com/news/2008/jun/17/offered-more-money-homeowners-respond-taking-out-m/.

Teal, Louise. *Breaking into the Current: Boatwomen of the Grand Canyon*. U of Arizona P, 1994.

"The Ten Tribes Partnership." Tentribespartnership, 2020, tentribespartnership.org/.

Teorey, Matthew. "Ecological Discourse in Craig Childs's *The Secret Knowledge of Water*." *The Journal of Ecocriticism*, vol. 2, no. 2, 2010, pp. 1–13, ojs.unbc.ca/index.php/joe/article/view/129.

Tompkins, Jane. *West of Everything: The Inner Life of Westerns*. Oxford UP, 1992.

Trava, Jose. "Sharing Water with the Colossus of the North." *Western Water Made Simple*, edited by Ed Marston, Island Press, 1987, pp. 171–81.

"Tribal Resources." *Glen Canyon Dam Adaptive Management Program Wiki*. 31 Mar. 2021, gcdamp.com/index.php?title=Tribal_Resources.

Trimble, Stephen, editor. *Red Rock Stories: Three Generations of Writers Speak on Behalf of Utah's Public Lands*. Torrey House Press, 2017.

Trimble, Stephen, and Terry Tempest Williams, editors. "An Act of Faith." *Testimony: Writers of the West Speak On Behalf of Utah Wilderness*. Milkweed, 1996, pp. 3–7.

True, Allen Tupper. *Tomorrow's Wisdom*. 1940, Colorado State Capitol, Denver. Photo by Kristal Kraft.

Trujillo, Leonard. "Interview with Leonard Trujillo." *Oral Histories*, Colorado River Water Users Association, 28 July 2006, www.crwua.org/documents/about-us/oral-histories/Leonard-Trujillo.pdf.

Tucker, Mary Evelyn, and John Grim. "Overview of World Religions and Ecology." Forum for Religion and Ecology at Yale, 2009, fore.yale.edu/about-us/overview-of-world-religions-and-ecology/.

Turner, Christi. "How We Export Our Water to Asia." *High Country News*, 2 May 2014, www.hcn.org/issues/46.7/how-we-export-our-water-to-asia.

Turner, Frederick Jackson. 1920. *The Frontier in American History*. U of Arizona P, 1986.

Udall, Bradley, and Jonathan Overpeck. "The Twenty-First Century Colorado Hot Drought and Implications for the Future." *Water Resources Research*, vol. 53, no. 3, Mar. 2017, pp. 2404–18. AGU, doi.org/10.1002/2016WR019638.

Udall, Stewart. "August 23, 1962, Letter to Senator Anderson." Box 31, Folder 22. MSS 20 BC Clinton P. Anderson Papers, University of New Mexico, Center for Southwest Research, Albuquerque.

———. "Remarks of Secretary of the Interior Stewart L. Udall, at the Dedication of Navajo Dam, San Juan County, New Mexico, September 15, 1962." Box 65, Folder 30. MSS 386 BC Joseph M. Montoya Papers, University of New Mexico, Center for Southwest Research, Albuquerque.

United States, Congress, House, Committee of Interstate and Foreign Commerce. *Renaming of the Grand River, Colorado*. Government Publishing Office, 18 Feb. 1921. Google Books, www.google.com/books/edition/Renaming_of_the_Grand_River_Colorado_Hea/0g22-Fsn75gC?hl=en&gbpv=1.

"Upper Colorado River Basin Compact of 1948." *Law of the River*, Lower Colorado Region, Bureau of Reclamation, www.usbr.gov/lc/region/pao/lawofrvr.html#ucbasin.

"U.S. Senate Passes Navajo Utah Water Rights Settlement Act." *Native News Online*, 9 June 2020, nativenewsonline.net/currents/u-s-senate-passes-navajo-utah-water-rights-settlement-act.

"Utah's Regional M&I Water Conservation Goals." *Regional Conservation Goals*. Utah Department of Water Resources, Nov. 2019, conservewater.utah.gov/regional-water-conservation-goals/.

Valentine, H. M. "Navajo Trail Association Inc." Box 31, Folder 22. MSS 20 BC Clinton P. Anderson Papers, University of New Mexico, Center for Southwest Research, Albuquerque.

"The Voices of Grand Canyon." Grand Canyon Trust, 20 Feb. 2020, www.grandcanyontrust.org/native-american-stories-grand-canyon.

Wald, Sara D., et al. *Latinx Environmentalisms: Place, Justice, and the Decolonial*. Temple UP, 2019. ProQuest, ebookcentral.proquest.com/lib/usd/reader.action?docID=5915565.

Walka, Ann Weiler. *Waterlines: Journeys on a Desert River*. Red Lake Books, 1993.

Walls, Laura Dassow. "Seeking Common Ground." *Coming into Contact: Explorations in Ecocritical Theory and Practice*, edited by Annie Merrill Ingram et al., U of Georgia P, 2007, pp. 199–208.

Ward, Evan R. *Border Oasis: Water and the Political Ecology of the Colorado River Delta, 1940–1975*. U of Arizona P, 2003.

"Water 2025: Preventing Crises and Conflict in the West." 2003. J. "Chips" Barry III Papers. Water Resources Archive, Colorado State University, Fort Collins. hdl.handle.net/10217/90221.

Waterman, Jonathan. *Running Dry: A Journey from Source to Sea Down the Colorado River*. National Geographic, 2010.

Waters, Frank. *People of the Valley*. 1941. Swallow Press, 1969.

Webb, Roy. *Call of the Colorado*. U of Idaho P, 1994.

Weiser, Matt. "Utah, America's Thirstiest State, Wrestles with Unmetered Water Use." *NewsDeeply*, www.newsdeeply.com/water/articles/2018/03/05/utah-americas-thirstiest-state-wrestles-with-unmetered-water-use.

Wharton, James. *Utah: The Land of Blossoming Valleys*. Colonial Press, 1922.

What the Colorado River Storage Project Means to the Nation's Largest Indian Tribe: The Navajos. Box 29, Folder 16. MSS 146 Arthur V. Watkins Papers. L. Tom Perry Special Collections, Brigham Young University, Provo, Utah.

White, Lynn, Jr. "The Historical Roots of Our Ecological Crisis." *Science*, vol. 155, no. 3767, 10 Mar. 1967, pp. 1202–7. JSTOR, www.jstor.org/stable/1720120.

Whyte, Kyle Powys. "Indigenous Experience, Environmental Justice, and Settler Colonialism." *SSRN*, 25 Apr. 2016, dx.doi.org/10.2139/ssrn.2770058.

Widtsoe, John A. *Don't Waste Water*. Box 3, Folder 4. ACCN 1562 John Andreas Widtsoe Papers. Special Collections and Archives, University of Utah, Salt Lake City.

———. *Dry Farming: A System of Agriculture for Countries Under Low Rainfall*. 1910. Macmillan, 1920.

———. *In a Sunlit Land*. Deseret News Press, 1952.

———. *A Journal of John A. Widtsoe*, edited by A. R. Mortensen. *Utah Historical Quarterly*, reprint from vol. 23, July 1955. L. Tom Perry Special Collections, Brigham Young University, Provo, Utah.

———. *The Principles of Irrigation Practice*. Macmillan, 1914.

———. *Success on Irrigation Projects*. John Wiley & Sons, 1928.

Wild, Peter. *Desert Literature: The Middle Period*. Boise State University, 1999.

———. Introduction. *The Desert: Further Studies in Natural Appearances*, by John C. Van Dyke, Johns Hopkins UP, 1999.

———. *The Opal Desert: Explorations of Fantasy and Reality in the American Southwest*. U of Texas P, 1999.

Wilkinson, Charles F. *Crossing the Next Meridian: Land, Water, and the Future of the West*. Island Press, 1992.

———. *Fire on the Plateau: Conflict and Endurance in the American Southwest*. Island Press, 2004.

Williams, A. Park, et al. "Large Contribution from Anthropogenic Warming to an Emerging North American Megadrought." *Science*, vol. 268, no. 6448, 17 Apr. 2020, pp. 314–18. AAAS, doi: 10.1126/science.aaz9600.

Williams, Terry Tempest. *Red: Passion and Patience in the Desert*. Pantheon, 2001.
Wister, Owen. Foreword. *Through the Grand Canyon from Wyoming to Mexico*, by Ellsworth L. Kolb, 1912. Macmillan, 1952, pp. ix–xii.
Worster, Donald. *Nature's Economy: A History of Ecological Ideas*. 2nd ed., Cambridge UP, 1994.
———. *A River Running West: The Life of John Wesley Powell*. Oxford UP, 2001.
———. *Rivers of Empire: Water, Aridity, and the Growth of the American West*. Oxford UP, 1985.
———. *Under Western Skies: Nature and History in the American West*. Oxford UP, 1992.
———. *The Wealth of Nature: Environmental History and the Ecological Imagination*. Oxford UP, 1994.
Yates, Norris W. *Genre and Gender: An Introduction to Women Writers of Formula Westerns, 1900–1950*. U of New Mexico P, 1995.
Ybarra, Pricilla. "Erasure by U.S. Legislation: Ruiz de Burton's Nineteenth-Century Novels and the Lost Archive of Mexican American Environmental Knowledge." *Environmental Criticism for the Twenty-First Century*, edited by Stephanie LeMenager, Teresa Shewry, and Ken Hiltner, Routledge, 2011, pp. 135–47.
———. *Writing the Goodlife: Mexican American Literature and the Environment*. U of Arizona P, 2016.
Yo Soy Rojo El Rio Colorado. Produced by Pete McBride, American Rivers, 22 Apr. 2015, americanrivers.org/rivers/films/yo-soy-rojo-el-rio-colorado/.
Zamora-Arroyo, Francisco, et al. "Collaboration in Mexico: Renewed Hope for the Colorado River Delta." *Nevada Law Journal*, vol. 8, no. 3, 2008, pp. 871–89. Scholarly Commons @ UNLV Law, scholars.law.unlv.edu/nlj/vol8/iss3/7/.
Zepeda, Ofelia. "Ḍ'ac 'O'odham" / "We Are Papago." *When It Rains: Tohono O'odham and Pima Poetry*. 1982. Edited by Ofelia Zepeda, U of Arizona P, 2019, pp. 74–75. Project Muse, muse.jhu.edu/book/63641.
———. *Ocean Power: Poems from the Desert*. U of Arizona P, 1995.
Zielinski, Sarah. "The Colorado River Runs Dry." *Smithsonian Magazine*, Oct. 2010, www.smithsonianmag.com/science-nature/the-colorado-river-runs-dry-61427169/.
Zwinger, Ann H. *Downcanyon: A Naturalist Explores the Colorado River Through the Grand Canyon*. U of Arizona P, 1995.
———. *Run, River, Run: A Naturalist's Journey Down One of the Great Rivers of the American West*. U of Arizona P, 1975.

Index

Page numbers in *italics* indicate illustrations.

About the Author

PAUL FORMISANO is associate professor of English and director of writing at the University of South Dakota. His teaching and research address sustainability issues in the American West.